# Biotic and Abiotic Stress Tolerance in Plants

# Biotic and Abiotic Stress Tolerance in Plants

Barry Day

MURPHY & MOORE
www.murphy-moorepublishing.com

**Murphy & Moore Publishing,**
1 Rockefeller Plaza,
New York City, NY 10020, USA

Visit us on the World Wide Web at:
www.murphy-moorepublishing.com



ISBN: 978-1-63987-080-6 (Hardback)



**Cataloging-in-Publication Data**

Biotic and abiotic stress tolerance in plants / Barry Day.
p. cm.
Includes bibliographical references and index.
ISBN 978-1-63987-080-6
1. Plants--Effect of stress on. 2. Plants--Effect of stress on--Genetic aspects.
3. Stress (Physiology). I. Day, Barry.
SB112.5 .B56 2021
632.1--dc23

# TABLE OF CONTENTS

# PREFACE

This book is a culmination of my many years of practice in this field. I attribute the success of this book to my support group. I would like to thank my parents who have showered me with unconditional love and support and my peers and professors for their constant guidance.

Plants experience biotic and abiotic stresses. The negative impact that other living organisms have on plants causes biotic stress. Some of these stressors are viruses, parasites, bacteria and weeds. Abiotic stress refers to the harmful impact of non-living factors such as wind, temperature, floods, droughts, sunlight and salinity. Plants deal with a variety of these biotic and abiotic stressors. Such stress can be detrimental as well as beneficial for the plants. The tolerance to these stressors depend upon the climate in which the plant grows and the ability of the plant to resist these stresses. Biotic stress affects the process of photosynthesis in plants whereas abiotic stress is known to affect the roots of the plants. This book provides comprehensive insights into stress tolerance in plants. It also unfolds the innovative aspects related to this field which will be crucial for the holistic understanding of the subject matter. This book aims to serve as a resource guide for students and experts alike and contribute to the growth of the discipline.

The details of chapters are provided below for a progressive learning:

Chapter – Biotic Stress in Plants

Biotic stress is defined as a stress that is caused in plants due to damage instigated by other living organisms, including fungi, bacteria, viruses, parasites, weeds, insects, and other native or cultivated plants. This is an introductory chapter which will briefly introduce about biotic stress in plants.

Chapter – Abiotic Stress in Plants

Abiotic stress is defined as the negative impact of non-living factors on living organisms in a specific environment. The major cause of poor plant growth and reduced crop yields includes abiotic stresses, such as drought, salinity, low and high temperatures. This chapter has been carefully written to provide an easy understanding of abiotic stress in plants.

Chapter – Types of Biotic and Abiotic Stress

There are many types of biotic and abiotic stress such as chilling stress, heat stress, moisture stress, flood stress, drought stress, mineral stress, oxidative stress, ultraviolet radiation stress, salinity stress, etc. The topics elaborated in this chapter will help in gaining a better perspective of these types of biotic and abiotic stress in plants.

Chapter – Factors affecting Stress Tolerance

There are diverse factors affecting stress tolerance that are studied under the domain of biotic and abiotic stress in plants. It includes transcription factors, role of fungal endophytes, soil bacteria and endophytic microbes against biotic and abiotic stress tolerance. This chapter discusses the aspects related to various factors affecting stress tolerance in detail.

**Barry Day**

# Biotic Stress in Plants

CHAPTER 1

Biotic stress is defined as a stress that is caused in plants due to damage instigated by other living organisms, including fungi, bacteria, viruses, parasites, weeds, insects, and other native or cultivated plants. This is an introductory chapter which will briefly introduce about biotic stress in plants.

Biotic stress arises from the other living organisms that the plants coexist and interact with. Biotic stress factors such as pathogens and herbivores can threaten food security and result in heavy monetary loses.

These stressors can be detected by visual field observations of symptoms, and the presence of causal factors can also be detected through field observations or through laboratory examinations.

## Pathogens

Pathogens are organisms that cause disease in plants and can be viruses, bacteria, protists, fungi, or nematodes. They can attack all the parts of the plant, affecting plant growth in myriad ways and reducing the productivity of crops. They can be transmitted by insects, wind, water, and even people.

## Herbivores

Herbivores can be small pests or mammals which feed on the leaves and other parts of the plant and can even damage entire crops in severe cases.

- Small pests include insects, mites, snails, slugs, and all kinds of bugs. In the case of insects, caterpillars and mature insects can feed on plants. Large-scale outbreaks of pests such as locusts, or fruit flies can damage whole regions in extreme cases. In addition, many pests act as vectors of disease.
- Rodents do not cause just stress: They can eat an entire plant, or even entire crops, and have previously resulted in famines.
- Ungulates: Such as cows, goats, and sheep when allowed to graze uncontrolled, can cause varying levels of loss in biomass and threaten crop yield.

## Detecting Stress Factors

The effect of stress can influence the physiology of plants, change regulatory networks, and alter gene expression. It is often difficult to pinpoint the exact cause of stress in plants, as more than one factor is usually involved, and they will interact synergistically. For example, saline soils are also likely to be suffering from waterlogging, and this makes it difficult to separate the effects of the various stress factors. Therefore, a fair level of expertise is required to identify stress factors. The availability of field instruments for the detection of some factors makes non-invasive measurement easier, with the purpose of narrowing down stress factors without further stressing the plants.

A number of measurements instruments can be used to detect stress in plants, crops and forests. For example:

- Canopy analysis and measuring leaf area index and photosythentically active radiation using a plant canopy analyzer.
- Root analysis measuring root growth and changes using a root imager.
- Measurement of photosynthesis, respiration, transpiration, stomatal conductance, PAR and internal $CO_2$ using a photosynthesis analyzer.
- Measurement of chemical concentrations, color, photochemical reactions such as photosynthesis, and physical or optical properties such as film thickness, index of refraction, and extinction coefficient using a spectrometer.

# PLANT RESPONSE TO BIOTIC STRESS

Hypersensitive Response (HR): On being attacked by insects or a pathogenic microorganism, typically a plant responds with:

- Formation of pathogenesis-related (PR) proteins.
- Biosynthesis of phytoalexins.
- Changes in composition and physical properties of cell walls and finally.
- Process of programmed cell death wherein the cells immediately surrounding the infection site die rapidly to deprive the pathogen of nutrient supply and thus checking its spread in host plant and leaving a necrotic lesion at the site of invasion. All these responses are collectively known as hypersensitive response or reaction (HR).

A brief account of these follows:

Pathogenesis-Related (PR) Proteins: These are products of defense-related genes that are activated by microbial infection and include hydrolytic enzymes such as:

- Proteinase inhibitors which inhibit activities of proteolytic enzymes secreted by the pathogen.
- Lytic enzymes such as glucanases, chitinases, and other hydrolases that attack and degrade the cell walls of the pathogen.

Biosynthesis of Phytoalexins: The microbial infection also activates genes that encode enzymes for the synthesis of phytoalexins. Phytoalexins are a chemical diverse group of secondary metabolites (chiefly isoflavonoids and sesquiterpenes) with strong antimicrobial activity. The isoflavonoids medicarpin from alfalfa and glyceolin from soybean and sesquiterpenes such as rishitin from tomato and potato and capsidiol from tobacco and pepper are well known examples of phytoalexins.

Changes in composition and Physical properties of host cell walls: In response to pathogen invasion, lignin, callose, suberin and some hydroxy-proline rich glycoproteins are synthesized and accumulated in host cell walls to strengthen the latter and physically blocking the spread of the invading pathogen.

(Death (PCD Programmed Cell): The hypersensitive response culminates in rapid death of cells around the infection site depriving pathogen of the nutrient supply and limiting its spread in host plant and leaving necrotic lesions (small regions of dead tissues) at the site of invasion. The rest of the plant however, remains unaffected. Recent researches have shown that the hypersensitive response is preceded by accumulation of nitric oxide (NO) and active oxygen species (including the superoxide anion $O_2^-$, hydrogen peroxide $H_2O_2$ and hydroxyl radical OH).

The production of active oxygen species (known as the oxidative burst) and nitric oxide (a secondary messenger in many signalling pathways in animals and plants) appears to be prerequisite for activation of hypersensitive response. Induction of PCD is prevented in absence of any of these two signals.

It is believed that a plasma membrane located NADPH dependent oxidase reduces the $O_2$ to produce superoxide anions. The latter in turn are converted into hydrogen peroxide and hydroxyl radicals. The active oxygen species especially the hydroxyl radicals may contribute to PCD as part of the hypersensitive response or these may act to kill the pathogen directly.

A transient increase in cytosolic $Ca^{2+}$ concentration is required for the activity of the enzyme NO synthase which converts the amino acid arginine into nitric oxide.

Consequent to pathogen infection, both oxidative burst and NO production are activated by a transient change in plasma-membrane permeability resulting in influx of $H^+$ and $Ca^{2+}$ ions into the cell and efflux of $K^+$ and $CI^-$ ions.

## Mechanism of Recognition of the Potential Pathogen to Initiate Defence Response in Plants

Not all plants are resistant to disease caused by pathogens. Researchers have shown that resistance of plants to microbial pathogens has an underlying genetic basis. The pathogens carry avr genes (virulence genes) while the host plants carry corresponding resistance genes called R genes. Disease occurs when the pathogen lacks avr genes or the host plant does not carry dominant R genes.

The avr genes are believed to encode enzymes for the production of specific substances called elicitors while R genes encode protein receptors that recognise and bind with these elicitors to initiate the hypersensitive response.

The elicitors (L., elicere, to entice) are substances originating from pathogens that include proteins, peptides, sterols, and polysaccharide fragments arising from cell walls of pathogen or outer membrane or a secretion process. Sometimes, polysaccharide fragments resulting from initial degradation of the host plant cell walls by pathogen may elicit a hypersensitive response.

The receptor proteins encoded by R genes are located on plasma-membrane and have a leucine-rich domain which is repeated inexactly many times in the amino-acid sequence. These domains may extend outer or inner to the plasma-membrane and bind with specific elicitors to recognise the pathogen. (R genes comprise one of the biggest gene families in plants).

## Secondary Acquired Resistance (SAR)

The hypersensitive response described earlier is limited to near vicinity of the initial site of infection by pathogen. But, often the entire host plants develops increased resistance against pathogens over a period of time ranging from few hours to several days following initial infection at one site of the plant. This phenomenon wherein “a single encounter with the pathogen increases resistance to entire plant to future attacks by pathogens” is called as secondary acquired resistance (SAR) and is part of the plant’s immune response.

SAR appears to result from increased levels of some secondary metabolites and other defense compounds such as chitinases and other hydrolytic enzymes. However, the mechanism of SAR is not clearly understood. One component of the signalling pathway is likely to be salicylic acid (SA) that is a benzoic acid derivative and a secondary metabolite. It has been shown in variety of plants that the infection by pathogen results in increased levels of SA in the zone of infection that establishes SAR in distant regions of thc plant.

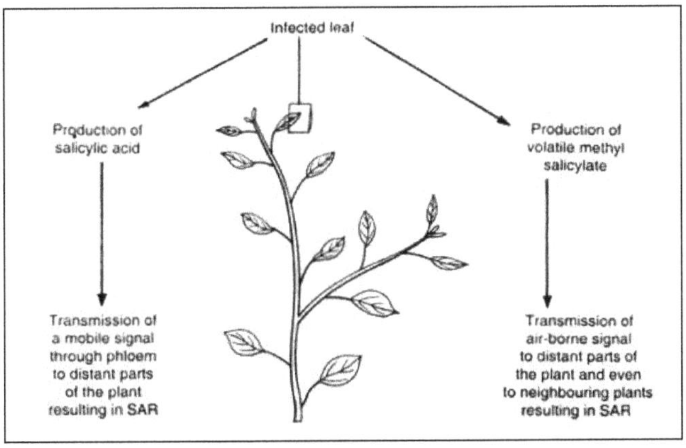


Possible roles of salicylic acid and methyl salicylate in SAR.

The transmission of SAR signal from infection site to other parts of the plant is very rapid (3 cm/h) and possibly occurs through vascular tissue. Phloem is now considered to be the pathway of SAR signal.

Salicylic acid is not the mobile SAR signal. Maldonado et al have shown that in Arabidopsis, mutations in DIR1 gene (Defective in Induced Resistance 1 gene) inhibit SAR response. This gene is specifically expressed in phloem and encodes a lipid-transfer protein and it has been suggested that the long distance SAR signal might be a substance derived from a lipid.

Apart from phloem-mobile signals, the plant may develop SAR through air-borne signals. Salicylic acid may be converted into its methyl ester, methyl salicylate that is a moderately volatile substance. (Structures of Salicylic acid and methyl salicylate are given in figure). Methyl salicylate may function as volatile air-borne SAR-inducing signal that is transmitted to distant non-infected parts of the plant and even to non-infected neighbouring plants making them resistant to pathogens attack.

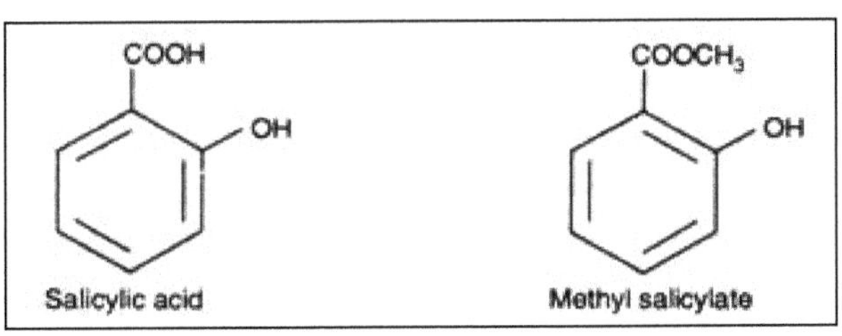


Structure of salicylic acid methyl salicylate.

## Advances in Plant Tolerance to Biotic Stresses

Plant mechanisms of resistance to various pathogens and insect pests are known to involve an array of morphological, genetic, biochemical and molecular processes. These mechanisms may be expressed continuously (constitutively) as preformed resistance, or they may be inducible and deployed only after attack. Plant success in deploying these resistance mechanisms is an evolved ability to persist in unfavourable and variable environments. The recent realization that plant mechanisms of disease/insect resistance or susceptibility are related to mechanistic animal immunity has significantly reshaped our view of plant immunity. The identification of plant pattern recognition receptors (PRRs) that sense pathogens' or insect pests' conserved molecules termed pathogen-associated molecular patterns or herbivore-associated molecular patterns (PAMPs/MAMPs/HAMPs)—and the subsequent PAMP-triggered immunity (PTI) is a paradigm for plant-pathogen interaction studies.

On the other hand, the ability of pathogens/insect pests to suppress or evade PTI, as a structural and functional basis of pathogen survival and evolutionary dynamics in their feeding mechanisms has revitalized research on the so-called 'gene-for-gene' effector induced resistance in plants. It is now clear that effectors are important determinants of pathogens' ability to evade the plant's arsenal targeted towards PAMPs/HAMPs. Effector induced resistance or vertical resistance, often interchangeably translated in modern terms as effector triggered immunity (ETI), is the most successful means of controlling pathogens able to evade PTI. ETI engages a compensatory mechanism within the defense network to transcriptionally coordinate and boost the defense output against pathogens. ETI mostly relies on the endogenious NB-LRR protein products encoded by the resistance (R)-genes. Although R gene mediated resistance is generally not durable, ETI is now effectively deployed through pyramiding of several resistance (R)-genes in the same cultivar, which increases resistance durability and spectrum.

Another aspect of resistance that has gained significance in plant defence studies is the systemic acquired resistance (SAR), in which defence proteins accumulate not only at the site of infection but also systemically in uninfected tissues and plants. SAR provides long-term defense against a broad-spectrum of pathogens and insects. Another form of induced resistance, which, in many aspects, is similar to SAR, is induced systemic resistance (ISR). ISR is potentiated by plant growth promoting rhizobacteria (PGPR), many of them belonging to Pseudomonas species. Obviously, the sessile nature of plants requires an efficient signalling system capable of detecting, transporting and interpreting signals produced at the plant-pathogen interface, and SAR and ISR provide a practical means to confer a fitness advantage to plants in conditions of high disease pressure, since plants are primed to more quickly and effectively activate their defences ahead of pathogen/insect attack. Plants also defend themselves through RNA interference to target and inactivate invading nucleic acids from viruses, and more recently fungal pathogens.

## Plant Defence Mechanisms in Response to Pathogens

Plants respond to various pathogens through an intricate and dynamic defence system. The mechanism of defence has been classified as innate and systemic plant response. An innate defence is exhibited by the plant in two ways, viz., specific (cultivar/pathogen race specific) and non-specific (non-host or general resistance). The molecular basis of non-host resistance is not well studied, but presumably relies on both constitutive barriers and inducible responses that involve a large array of proteins and other organic molecules produced prior to infection or during pathogen attack Constitutive defences include morphological and structural barriers (cell walls, epidermis layer, trichomes, thorns, etc.), chemical compounds (metabolites, phenolics, nitrogen compounds, saponins, terpenoids, steroids and glucosinolates), and proteins and enzymes. These compounds confer tolerance or resistance to biotic stresses by not only protecting the plant from invasion, but also giving the plant strength and rigidity. The inducible defences, e.g., the production of toxic chemicals, pathogen-degrading enzymes e.g., chitinases and glucanases, and deliberate cell suicide are conservatively used by plants because of the high energy costs and nutrient requirements associated with their production and maintenance. These compounds may be present in their biologically active forms or stored as inactive precursors that are converted to their active forms by host enzymes in response to pathogen attack or tissue damage. Plant defence strategies involving these compounds can fall in either category, innate or SAR. Although innate immunity is of greater efficiency and is the most common form of plant resistance to microbes, both defence strategies depend on the ability of the plant to distinguish between self and non-self molecules.

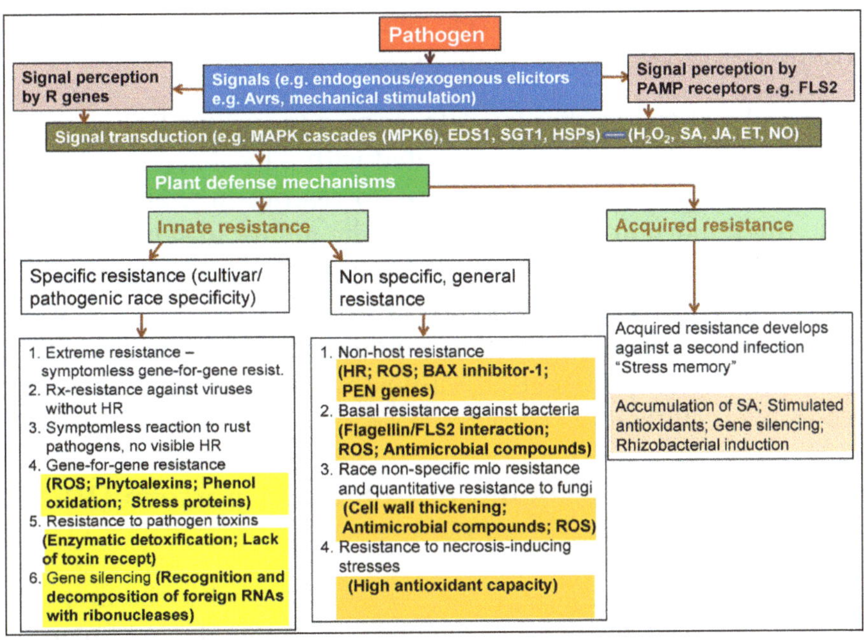


Overview of cellular mechanisms of biotic stress response leading to innate immunity and systemic acquired resistance. Plant PRRs or R-genes perceive PAMPS/DAMPs and effectors, respectively. Inside the cell, an overlapping set of downstream immune responses results from the PTI/ETI continuum. This includes the activation of multiple signaling pathways involving reactive oxygen species (ROS), defense hormones (such as salicylic acid, jasmonic acid and ethylene), mitogen activated protein kinases (MAPK), and transcription factor families, e.g., AP2/ERF, WRKY, MYB, bZIP etc. these signals activate either innate response or acquired immune response or both.

## Innate Immunity

Innate immunity in plants is divided into microbial-associated molecular-pattern-triggered immunity (MTI; also called PTI) and effector-triggered immunity (ETI). In MTI/PTI, innate immunity is defined by receptors for microbe-associated molecules, conserved mitogen-associated protein kinase signalling cascades and the production of antimicrobial peptides/compounds. Recognition of microbes is divided into two branches, one involving slowly evolving microbial or pathogen-associated molecular patterns, such as fungal chitin, xylanase or bacterial flagellin, lipopolysaccharides and peptidoglycans, and the other that responds to a compromised 'self', also called damage-associated molecular patterns (DAMPs). Both PAMPs and DAMPs are recognized by trans membrane pattern recognition receptors (PRRs).

A common strategy employed by adapted pathogens is to secrete effector proteins that avoid or regulate PTI recognition. To counter this stealth afforded by the microbial effectors, plants have evolved an intracellular surveillance involving polymorphic NB-LRR protein products encoded by resistance (R) genes, named after their characteristic feature due to the presence of nucleotide binding (NB) and leucine-rich repeat (LRR) domains. This type of plant defence is referred to as ETI and is synonymous to pathogen race/host plant cultivar-specific plant disease resistance.

Generally, PTI and ETI trigger similar defence responses, but ETI is much faster and quantitatively stronger. ETI is often associated with a localized cell death termed the hypersensitive response (HR) that functions to restrict further spread of microbial attack. Hence, the important feature of ETI is the ability to sense microbe-mediated modifications inferred on points of vulnerability in the host, whereas PTI is able to sense infectious-self and non-self. By guarding against weak points or even setting up decoys to confuse invaders, ETI is an efficient defence system for more progressed infections, whereas PTI is important for non-host resistance and for basal immunity in susceptible host plant cultivars.

## Pathogen or Microbial-associated Molecular-pattern (PAMP/MAMP)-triggered Immunity (PTI)

PTI (formerly called basal or horizontal disease resistance) is the first facet of active plant defence and can be considered as the primary driving force of plant-microbe interactions. PTI involves the recognition of conserved, indispensable microbial elicitors known as PAMPs by PRRs of either the receptor-like kinase (RLK) or receptor-like proteins (RLPs) families, which are membranous bound extracellular receptors. RLPs resemble the extracellular domains of RLKs, but lack the cytosolic signalling domain, whereas RLKs have both extracellular and intracellular kinase domains. Instances of hetero-oligomeric complexes between RLKs and RLPs have been reported to occur, and to complement each other in PAMP detection, Examples of RLPs include the S locus glycoprotein (SLG), CLAVATA2 and Xa21D. RLKs are numerous, despite different configurations, both RLKs and RLPs receptors contribute to blocking infection before the microbe gains a hold on the plant.

PAMPs occur throughout the pathogen classes, including bacterial flagellin and EF-Tu (elf18), fungal chitin (CEBiP) and mannans of yeast, xylanase (LeEIX1/2) and Oomycetes' heptaglucan (HG). The early responses induced by PAMPs occur within minutes to hours and are varied, ranging from rapid ion fluxes across the plasma membrane, oxidative burst, activation of mitogen-activated protein kinases (MAPKs) and calcium-dependent protein kinases (CDPKs) to local induction of defence-related genes or pathogen cell wall/cell membranes lyasing enzymes/peptides, e.g., chitinases, glucanases and defensins. Other responses may include production of antimicrobial phytoalexins, plant cell wall modifications, e.g. deposition of papillae, enriched with (1,3)-β-glucan cell wall polymer, callose, lignin biosynthesis, or changes in cell wall proteins and pectic polysaccharide structures. When the pathogen gains entry and initiates colonization, a concerted effort of both PTI and ETI may be required to restrict further colonization. In the event that ETI is not active, PTI could probably contribute to effective plant resistance as much as ETI, if the capacity to recognize undetected epitopes could be engineered into plants.

## Specific Examples of PTI in Plants

## Flagellin-induced Resistance

Flagellin constitutes the main building block of bacterial flagellum, and is so far the best characterized PAMP in plants. A 22 amino acid peptide-spanning region in the N-terminal part of flagellin of Pseudomonas syringae is sufficient to elicit the whole array of typical immune responses in a broad variety of plants. The PRR responsible for flagellin perception in the model plant Arabidopsis thaliana is the leucine-rich repeat receptor-like kinase (LRR-RLK) FLAGELLIN-SENSING2 (FLS2). Functional FLS2 homologs have been identified in other major groups of higher plants, including tomato, grapevine, Nicotiana benthamiana and rice, suggesting that the receptors for the flg22 epitope of bacterial flagellin are evolutionarily ancient and conserved. Despite evolutionary conservation, FLS2 proteins from different plant species, such as tomato flagellin receptor (LeFLS2), grapevine (VvFLS2) and A. thaliana (AtFLS2), still exhibit different perception specificities to elicitation determinants of flagellins. This suggests that the domains found in FLS may have undergone some functional innovations that contribute to different perception specificities. Flagellin also seems to be recognized by other means in certain plant species. For instance, in rice, flg22 epitope does not allow the activation of PRR, but flagellin induces cell death. Moreover, the glycosylation status of flagellin proteins is emerging as a determinant of recognizing adapted and non-adapted bacteria by Solanaceae plants, such as tobacco and tomato. More recently, another flagellin, flgII-28, was identified in Solanaceae, though the corresponding PRR is yet to be identified. Both flg22 and flgII-28 are physically linked by a stretch of 33 amino acid residues, suggesting that both molecules are detected by the same receptor, FLS2.

The signalling events triggered in plant cells following flg22 detection include rapid binding of FLS2 to BAK1 (BRI1-associated kinase 1) by reciprocal transphosphorylation of their kinase domains. The plasma membrane localized receptor-like cytoplasmic kinase BOTRYTIS-INDUCED KINASE 1 (BIK1) and related PBS1-LIKE (PBL) kinases associate with FLS2/BAK1. The complex formed triggers multiple rapid phosphorylation events resulting in BIK1 release. BIK1 plays a central role in conveying signals from not only FLS2 but also other PRRs, including EFR, CERK1 and the DAMP receptor, PEPR1/PEPR2. The signal transduction downstream of flg22 perception includes a $Ca^{2+}$ burst, activation of CDPKs and RbohD required for the ROS burst and induction of MAPK cascades. These signalling cascades activate transcriptional reprogrammers such as the WRKY TFs, which are required for induction of defence genes.

## Elongation Factor (EF-Tu) Induced Resistance

Elongation factor Tu (EF-Tu) is the most abundant bacterial protein originally isolated from Escherichia coli, and acts as PAMP in Brassicaceae family members including A. thaliana. The conserved N-acetylated epitope elf18 (first 18 amino acids of the protein) is sufficient to trigger defence responses in plants. The shorter peptide, elf12 (first 12 N-terminal amino acids), comprising the acetyl group, is inactive as an elicitor but acts as a specific antagonist for EF-Tu–related elicitors. EF-Tu is recognized by the LRR-RLK EF-TU RECEPTOR (EFR) of the same subfamily (LRRXII) as FLS2. Interestingly, the ability to perceive elf18 epitope seems restricted to the plant family Brassicaceae. However, heterologous expression of EFR in the Solanaceae family, e.g., N. benthamiana and Solanum lycopersicum, makes them more resistant to a range of phytopathogenic bacteria, suggesting that EFR can be as well used to engineer broad-spectrum disease resistance in other families. More recently, EFa50 central region comprising Lys176 to Gly225 was found to be fully active as a PAMP in rice and induced $H_2O_2$ generation and callose deposition. Moreover, AtEFR-transformed rice plants were shown to be well responsive to the Xanthomonas oryzae derived elf18 peptide by strongly inducing ROS burst and expression of OsPBZ1 in transgenic cell cultures, further suggesting that EFR confers stable resistance across plant families.

The mechanism of EFR resistance is mediated by heteromeric complex formation. For instance, in rice, the complex formed between SOMATIC EMBRYOGENESIS RECEPTOR KINASEs (OsSERK2; an ortholog of BAK1) and XA21 binding protein 24 (XB24) is the most important component of XA21-mediated defence response. Four SERK co-receptor-like kinases interact with EFR within seconds to minutes of ligand binding, and once the ligand is perceived, EFR is rapidly phosphorylated, which triggers downstream signal activation, including the activation and release of BIK1. BIK1 plays a central role in conveying signals. Interaction between EFR and SERK also triggers the activation and release of other members of the cytoplasmic receptor-like kinase subfamily VII from the complex. Downstream components of these responses include activation of a RING finger ubiquitin ligase (XB3), MAPKs, the plant-specific ankyrin-repeat (PANK) containing protein XB25, and WRKY TFs.

Notwithstanding the FLS2 and EFR PRRs identified so far, relatively fewer PRR genes have been utilized to enhance plant resistance to bacterial pathogens through breeding and transgenic approaches, except a few that have been shown to be better adapted to defence signalling. The most famous example is that of Xa21 gene transferred from Oryza longistaminata, which confers high resistance to X. oryzae in rice. Heterologous expression of XA21 in Citrus sinensis, Lycopersicon esculentum and banana (Musa sp.) also conferred moderate resistance to Xanthomonas axonopodis pv. citri and resistance to Ralstonia solanacearum and Xanthomonas campestris pv. malvacearum in experiments under controlled conditions. The tomato RLP Ve1, which recognizes Ave1 from Verticillium dahliae race 1 is another inter-class example that confers stable resistance when transferred and expressed in Arabidopsis for use as a model genetic system. Taken together, XA21 and Ve1 are an example of engineered resistance strategy under controlled conditions, despite their taxonomic restrictions. However, more PRRs recognizing conserved molecular signatures in bacteria will need to be discovered and their complex interaction with the plant's physiology and metabolism and the environment understood, if the ambition of improving crop plants through genetic engineering of broad-spectrum disease resistance by gene transfer is to become more convincing.

## Plant Perception of PAMPs from Fungi and Oomycetes

Chitin, a homopolymer of β-(1,4)-linked N-acetylglucosamine (GlcNAc) unit, is a major constituent of fungal cell walls and is a classical PAMP. Chitin is an ideal point of attack during plant defence responses since glucosamine polymers are not found in plants. Upon pathogen contact with the host, plant chitinases (hydrolytic enzymes) break down microbial chitin polymers. Interestingly, different plants have evolved mechanisms that employ common factors for chitin perception, and this could be probably the reason for the evolution of pathogen counter measures, e.g., in the biotrophic fungal pathogen Cladosporium fulvum. In this context, the reaction of tomato with induction of defense-related, signal transduction and transcription genes to external chitin application supports the role of the described mechanisms.

The first chitin-binding PRR was identified in rice as the lysine motif (LysM)-RLP, and was named chitin-elicitor binding protein (CEBiP). CEBiP is a glycoprotein that localizes in the plasma membrane. Upon chitin binding, CEBiP homodimerizes and forms a hetero-oligomeric complex with the Chitin Elicitor Receptor Kinase 1 (OsCERK1), the rice ortholog of Arabidopsis AtCERK1. The binding thus forms a sandwich-type receptor system for chitin as described in. The mechanism of perception, however, varies between plant species. For example, AtCERK1 does not seem to employ CEBiP-like LysM-RLPs to induce typical immune responses such as reactive oxygen species and immune gene expression upon chitin perception. Instead, AtCERK1 binds directly to octamers of chitin, which in turn induce AtCERK1 homodimerization and the resultant immune signalling. Arabidopsis LysM (AtLYM2), the closest ortholog of AtCEBiP, and the rice LysM RLPs (OsLYP4 and OsLYP6) are also able to bind chitin. However, it is not clear whether AtLYM2/LYK4 also display the putative homodimerization induced by chitin perception. Two other orthologs of CEBiP, AtLYM1 and AtLYM3, which specifically bind PGN, but not chitin, interact with AtCERK1. This indicates that AtCERK1 is a multifaceted RLK that also forms hetero-oligomeric complexes with ligand-binding RLPs, probably across different plant families.

Fungal xylanases also function as fungal PAMPs by eliciting defence responses and promoting necrosis. In tomato, ethylene-inducing xylanases (EIXs) produced by Trichoderma species are perceived by two specific LRR-RLPs receptors, LeEix1 and LeEix2. Both receptors bind Eixs, but oLeEix2 is the primary mediator of defence responses. LeEix1 heterodimerizes with LeEix2 upon application of the Eixs and attenuates Eix-induced internalization and the subsequent signalling of the LeEix2 receptor. Microbial xyloglucan-specific endoglucanases (XEGs) have also been reported to induce plant defences. Fungal XEGs are inhibited by xyloglucan endoglucanase inhibiting proteins (XEGIPs), which so far have been characterized in tomato, carrot and tobacco.

Other PRRs that have been identified in plants in response to fungal PAMPs include the Brassica napus LepR3/Rlm2, for blackleg resistance, which perceives AVRLM1. In Arabidopsis, Rlm2 interacts with suppressor of BAK1-interacting receptor-like kinase 1 (AtSOBIR1), suggesting that SOBIR1 is a component of LRR-RLP-mediated resistance against Leptosphaeria maculans, which is similar to that formed by rice OsCERK1 and Arabidopsis AtCERK1. The tomato Cf proteins (Cf2, Cf4 and Cf9) that recognize the corresponding effector proteins (Avr2, Avr4 and Avr9) secreted by C. fulvum are other PRR-like receptors that were previously identified. Cf4 interacts with BAK1 in a manner similar to the rice ligand binding and associated receptor OsSERK/EFR.

Wheat and Arabidopsis RLP1.1 and RLP30 are also involved in antifungal defence, although the corresponding ligands are unknown so far. Several orphan PAMPs with unknown PRRs, from fungi or oomycetes that can trigger immune signalling have also been identified, including fungal ergosterol, oomycete arachidonic acid, elicitins (INF1), the transglutaminase-derived immunogenic epitope Pep13, cryptogein and cellulose-binding elicitor lectin (CBEL). Thus, further research is required to understand mechanistically how these orphan PAMPs are involved in PTI.

Taken together, the identification of several potential host plant receptor targets and receptor complexes, and their stability across plant species and in the field will greatly help to improve plant protection. Moreover, identification of several potential microbial molecules that act as PAMPs would increase chances of identifying more potential host plant PRRs for developing crops with higher resistance or inducible resistance.

## Plant Perception of Virus PAMPs

Although viral patterns inducing PTI are well known from animal systems, there is no similar pattern reported for plants. Instead, plant resistance to viruses is mediated by post-transcriptional gene silencing of viral RNA or ETI. Nevertheless, infection by compatible viruses can also induce defence responses similar to PTI. Typical PTI cellular responses in plant-virus interactions include ion fluxes, ROS production, ethylene, salicylic acid (SA), MAPK signalling and callose deposition. Commonly reported genes associated with PRRs in response to viruses include PEPs that encode longer peptides (ProPEP) from which small peptides (PEP) are derived. In Arabidopsis, AtPEP interact with two DAMP PRRs, PEP-receptor 1 (PEPR1) and PEPR2, both of which interact with BAK1 upon recognition of AtPEP. Thus, BAK1 is important for antiviral defence in Arabidopsis. Indeed, the bak1 mutants show enhanced susceptibility to three different RNA viruses (TMV-U1, ORMV and TCV) during compatible interactions. The immune response induced by PEPR-BAK1 interaction is a classical PTI. Another viral resistance mechanism, which is highly similar to BAK1 and BAK1-like Kinase 1 (BKK1), is exhibited by the viral nuclear shuttle protein (NSP)-interacting kinases (NIKs) from leucine-rich repeats containing receptor-like serine/threonine kinase (LRR-RKs) subfamily.

The ribonuclease III-type DICER-like (DCL) enzymes could be acting as PRRs perceiving viral nucleic acids and triggering immune responses equivalent to the zig-zag model first layer. The virus-derived molecules (e.g., dsRNAs) act as PAMPs, which trigger PTI and RNA interference (RNAi). However, PTI is typically a form of innate immunity, whereas RNAi induces a form of adaptive immunity. Thus, it is clear that a lot remains to be discovered to prove that virus-derived molecules trigger PTI.

## Plant Perception of Insect PAMPs

Molecular recognition via ligand-receptor binding phenomena is increasingly becoming important in insect-plant interactions. As reported earlier, the concept of PAMPS has been expanded to include herbivore-associated molecular patterns or damaged-self compounds produced after insect attack. HAMPs isolated and characterized to date include components found in insect oral secretions (proteins, fatty acid-amino acid conjugates (FACs), sulphur-containing fatty acids, as well as plant-derived molecules generated following insect herbivory, including degradation products of ATP synthase and cell walls. The insect oral secretion molecules are released by chewing insects and have been reported to induce ion imbalances, variations in membrane potential, changes in $Ca^{2+}$ fluxes and the generation of reactive oxygen species (ROS), which stimulate downstream signalling events in plants. $Ca^{2+}$ influx is obviously preceded by the opening of calcium channels, and it is likely that these channels are associated with plant receptors tuned to insect elicitors. Recently, a mechanism similar to PTI was reported in Arabidopsis in which LRR-RK BAK1 was shown to contribute to innate immunity against aphids. Moreover, application of synthetic FACs on wounded N. attenuate leaves strongly induced MAPK activity, and subsequently wound-induced modifications in the transcriptome, proteome and defensive secondary metabolites. Insect egg ovipositional fluids have also been shown to induce plant defences. Moreover, insect egg deposition on one leaf could induce volatile emission in the other egg-free leaves, suggesting that SAR could be involved after detection of insect eggs' associated molecules. An interesting example was reported in the oviposition by Pieris brassicae, which triggered SA accumulation and the subsequent induction of PAMP responsive gene expression associated with lectin-domain RK (LecRK), LecRK-I. Correspondingly, expression of the defence gene PR-1, which requires EDS1, SID2 and NPR1, was also detected, implicating the SA pathway downstream of the insect egg recognition.

Another mechanism that is closely related to the PAMP receptors in plant resistance to insects is the Mi-1 gene in tomato. The induction of Mi-1 confers resistance to Macrosiphum euphorbiae. A receptor-like kinase gene OsLecRK in rice, which confers basal resistance to Nilaparvata lugens, was recently suggested to be a PRR that recognizes molecules secreted by these insects. A similar mechanism was demonstrated in aphid infestation of Arabidopsis in which the immune response was apparently triggered by infiltration of aphid saliva. Consistent with this, infiltration of whole aphid extract from M. persicae was reported to activate PTI-like responses in Arabidopsis.

This notwithstanding, the insect HAMP-receptor binding phenomenon that allows plants to detect insects still remains less clear as to whether these responses are exclusively due to the specific perception of herbivores or due to different damage patterns or both.

## Infection Self-perception DAMPs

Plants can also sense self-molecules called damage-associated molecular patterns that are available for recognition only after cell/tissue damage. A perfect example is the Arabidopsis plasma membrane LRR receptor kinase (LRR-RK), designated PEPR1/PEPR2, which perceives AtPep peptides derived from propeptide (ProPEPs) encoded by a seven-member multigenic family (Pep1-Pep7). Both PEPR1 and PEPR2 were reported to be transcriptionally induced by wounding, treatment with methyl jasmonate, Pep peptides and pathogen-associated molecular patterns. Moreover, AtPep perception is part of a PTI amplification loop and is important for the induction of systemic immunity. In another example, hydroxyproline-containing glycopeptides (HypSys) and rapid alkalinization factor (RALF) peptides have been shown to induce an MAPK cascade in tomato cells. The precursors of HypSys and RALF are constitutively present in the plant cell walls. Microbial proteases or intracellular proteases release these peptides upon cell injury, making then to act as DAMPs.

Cell wall components derived from the enzymatic activity of highly specific microbial homogalacturonan (HGA) is another good example of DAMPs. The enhanced production of oligogalacturonic acid (OGA) fragments from plant cell walls potentially acts as DAMP, which are perceived by receptors such as RLK THESEUS1 (THE1), ER and WAK1. Plants may also rely on the recognition of cell wall degrading enzymes (CWDEs) by LRR-RLPs receptors, e.g., RBPG1 and LeEIX1-2. A decisive role of the composition and structure of plant cell wall polysaccharides, specifically of side chains of pectic polysaccharides, in elicitation of plant defence has also been described in tomato interaction with a bacterial pathogen, R. solanacearum. Thus, studying the expression of endogenous molecules and microbial cell wall degrading enzymes and their inhibitors, e.g., polygalacturonases (PGs) and polygalacturonase-inhibiting proteins (PGIPs) is a valuable approach to understanding the dynamics of plant-pathogen interactions as well as to develop a strategy to improve plant protection using induced plant endogenous molecules.

## Effector-triggered Immunity (ETI)

ETI (formerly called R-gene-mediated or vertical resistance) is based on the highly specific, direct or indirect interaction of pathogen effectors and the products of plant R genes according to the gene-for-gene theory. R genes encode proteins of the intracellular nucleotide-binding leucine-rich repeat (NB-LRR) class. The NB-LRR consist of N-terminal effector domain, central NB domain and C-terminal LRR domain, which largely vary in plants. Two major subgroups that have distinct N-terminal domains are generally recognized: (1) One group with a Toll–interleukin 1 receptor (TIR) domain are called TNLs, and (2) those with a coiled-coil (CC) domain are called CNLs.

In Arabidopsis, the CNLs functionally interact with the glycosylphosphatidylinositol (GPI) anchored protein—NON-RACE SPECIFIC DISEASE RESISTANCE 1 (NDR1), a positive regulator of SA accumulation, for signalling. Indeed, an ndr1 mutation compromises resistance conferred by the CC-NBS-LRR proteins RPS2, RPM1 or RPS5 to P. syringae expressing the avirulence effectors avrRpt2, avrB and avrRpm1, or avrPph3, respectively. In contrast, multiple TNLs functionally associate with ENHANCED DISEASE SUCEPTIBILITY 1 (EDS1) and PHYTOALEXIN DEFICIENT 4 (PAD4) for signalling. For instance, resistance conferred by the TIR–NBS–LRR protein RPS4, which recognizes avrRps4 in P. syringae is compromised in eds1 mutants. However, resistance mediated by some R genes is independent of EDS1/PAD4 and NDR1 or require additional co-activating proteins, suggesting existence of additional components for signal transmission during plant-pathogen interaction. Some of the regulatory components functionally associated with R genes for an effective HR mediated resistance include RAR1 (required for Mla12 resistance) and SGT1 (suppressor of the $G_2$ allele of skp1) proteins. RAR1 interacts with the N-terminal half of HSP90 that contains the ATPase domain. HSP90 also specifically interacts with SGT1 that contains a tetratricopeptide repeat motif and a domain with similarity to the co-chaperone p23. These observations suggest that R proteins require several co-activating proteins, although distinct downstream signalling pathways could be involved. There are also some NLRs containing N terminus other than the classical TIR and CC, either because their protein structures are not validated or due to lack of significant homology; they are referred to as non-TIR-type NLRs (nTNLs) or generally referred to as NLRs. Further work on non-sequenced genomes is likely to expand the number of NLRs, and probably refine functional difference associated with NLR repertoires.

Regardless of the NLR class, NB-ARC domain is the core nucleotide-binding fold in NB-LRR proteins. Four distinct subdomains constitute the NB-ARC domain, including nucleotide-binding (NB) fold and ARC1, -2 and -3 subdomains. ARC1 is a four-helix bundle, ARC2 is a winged-helix fold and ARC3 is a helical bundle. ARC1 and ARC2 are conserved in Caenorhabditis elegans CED-4, and plant NB-LRR R proteins, whereas

ARC3 is absent. Throughout the NB-ARC domain in R proteins, numerous conserved motifs (e.g., hhGRExE, Walker A or P-loop, Walker B, GxP, RNBS-A to D and MHD) have been reported. A mutation in these conserved motifs has shown their functional importance in the NB-LRR proteins, and is apparently a critical factor determining R gene functional effector recognition pattern differences. Generally, pathogen effector recognition by NLR and NLR expression are broadly characterized into (1) direct NLR-Effector interaction or (2) indirect NLR indirect surveillance of effector activities.

## Direct NLR-effector Interaction

NLRs maintain an ADP-binding inactive state in the absence of effectors. The binding of effectors induces conformational changes in NLRs, which allow ADP/ATP exchange. Consequently, the exchange of nucleotides triggers a second conformational change that activates the NB-LRRs' N-terminus (TIR or CC) to interact with and trigger downstream target processes. However, there is no substantial evidence on direct NLR-effector interaction that underlies resistance specificity in the NLR-effector combinations, apart from the yeast two-hybrid (Y2H) and in vitro interaction assays. A few examples that attempt to show the NLR-effector interaction include the Arabidopsis NLR RPP1 recognition of the oomycete effector ATR1 leading to Hyaloperonospora arabidopsidis (Hpa) resistance. Both the RPP1 receptor and ATR1 alleles from Hpa strains can be diverse. This diversity contributes to a spectrum of resistance phenotypes and effectors. For instance, the recognition specificity of RPP1-WsB (from the Wassilewskija ecotype) and RPP1-NdA (from the Niederzenz ecotype) vary. The RPP1-NdA recognizes a small subset of the ATR1 alleles recognized by RPP1-WsB, while the RPP1-WsB associates with the cognate Hpa effector protein, Atr1, through its LRR domain in a recognition-specific manner. Another example is the Arabidopsis NLR RRS1, a domain with sequence similarity to WRKY TFs, positioned after the LRR. The cognate effectors AvrRps4 and PopP2 directly interact with this WRKY-like domain to activate the downstream resistance components.

Together, the different R proteins have functional domains that can occupy different positions in NLRs. The functional domain positioning differences could be the reason behind several R genes that have been identified in plants. For instance, in rice more than 100 NLRs encoding genes have been described to confer resistance to strains of Magnaporthe oryzae. However, only few R proteins encoded by these genes have been characterized, which limits their deployment. A well-known structure for the recognition of M. oryzae effectors is that of AVR-Piz-t, which adopts a six-stranded β-sandwich structure and contains a single disulphide bond. The AVR-Pia and AVR1-CO39 have also been reported to be recognized by the R GENE ANALOGs (RGA4/RGA5) NLR pair through direct binding to a Heavy-Metal Associated domain (HMA; also known as RATX1) integrated into RGA5 after the LRR position. RGA4/RGA5 physically interact to prevent cell death mediated by RGA4 in the absence of AVR-Pia; the presence of the effector relieves this suppression, and induces cell death response, a mechanism that could also be described as indirect NLR surveilance. More recently, Maqbool et al.

also found that recognition of AVR-Pik by Pik is by direct binding to the HMA domain of Pik-1. However, the positioning of the HMA domain between the CC and NB-ARC region of Pik-1 and after the LRR in RGA5 is a striking difference between Pik-1 and RGA5. These conformational changes underlying direct effector binding could be causing immunity-related signalling differences. However, the intra and inter-molecular complexes mediating output may be conserved.

## Indirect NLR Surveillance of Effector Activities

During indirect recognition, the NLR guards the host protein by recognizing (monitoring) the modifications caused by the pathogen effector on the guarded protein. The guarded protein can either be the actual effector virulence target or a decoy inviting modification by the pathogen. An example of the indirect recognition of effectors by NLRs was demonstrated in the conserved Arabidopsis protein RPM1-interacting protein 4 (RIN4). RIN4 is targeted by multiple bacterial effectors, e.g., AvrRpt2, AvrRpm1 and AvrB, and is monitored for effector-induced modification by two plasma membrane CNL receptors, RPM1 (resistance to P. syringae pv. maculicola 1) and RPS2 (resistance to P. syringae 2). AvrB-induced phosphorylation and cis/trans isomerization coupled with conformational changes in RIN4 are sensed by RPM1 to activate immune signalling. AvrRpt2, being a cysteine protease, cleaves RIN4 and induces RIN4 degradation. In the absence of RPM1 and RPS2, RIN4 acts as a negative regulator of basal resistance, and in that capacity appears to be targeted for manipulation by multiple bacterial effectors.

The functioning of NLRs as genetically tightly linked pairs to deliver disease resistance was also recently reported. Moreover, Williams et al. demonstrated, by coupling crystal structure and functional analyses, that RPS4 and RESISTANT TO RALSTONIA SOLANACEARUM 1 (RRS1) TIR domains form homo and hetero-dimers through a common conserved interface that includes a core serine-histidine (SH) motif. Transient expression assays in tobacco revealed that the RPS4 TIR domain triggers an effector-independent cell death, which is dependent on the SH motif. Co-expression of the RRS1 TIR domain and RPS4 TIR impedes the auto-active cell death caused by RPS4 TIR, and this was found to be dependent on the RRS1 SH motif. This suggests that an inactive RRS1/RPS4 TIR hetero-dimer and the formation of an active RPS4 TIR homo-dimer compete to modulate signalling. Cesari et al. investigated the mode of action of RGA 4 and 5 that associate through their coiled-coil domains. RGA4 and RGA5 are tightly linked rice CC-NLRs, which functionally interact to modulate resistance to the rice pathogen M. oryzae. RGA5 modulates an effector independent cell death constitutively induced by RGA4 signalling. RGA5 domain on the C-terminus has a heavy-metal-associated domain, which is related to the cytoplasmic copper chaperone ATX1 from Saccharomyces cerevisiae (RATX1 domain). This domain is an AVR-Pia effector interacting domain in RGA5. Thus, the formation of the RGA4/RGA5 hetero-complex is crucial to regulate RGA4 activity in the absence of pathogen in rice. Hence, RGA4 acts as a signalling component regulated by its interaction with RGA5 that acts both as a repressor and a receptor that directly binds the AVR-Pia proteins. The apparent striking similarity between the RPS4/RRS1 and the RGA4/RGA5 functional models suggests that similarities are likely to be frequent between the different R genes present in dicots and monocots.

## Patterns of NLRs Signalling in Plant Defence

Most NLRs respond to the presence of proteins (effectors) delivered by adapted pathogens/parasites. Using suppressor screens, Gabriels et al., identified NRC1 (NLR protein required for HR-associated cell death 1) as a component of fungal resistance modulated by the tomato plasma membrane receptor-like resistance protein Cf-4 (C. fulvum 4). NRC1 mediates resistance and cell death induced by both membrane receptors and intracellular NLRs. This indicates that NRC1 is probably a downstream convergence point in ETI initiated at various cell locations. Indeed, silencing of NRC1 in N. benthamiana impairs the HR mediated by several other R proteins including two NLRs, Rx and Mi. Members of a conserved class of non-canonical CNLs also function in ETI, downstream of NLR effector recognition and have been designated as helper NLRs. Characterization of these non-canonical CNLs is required in order to track their interaction networks.

The downstream components of ETI signalling events partially overlap with PTI response, including activation of MAPK cascade and activation of TFs such as WRKYs. In Arabidopsis, three CNLs—activated disease resistance 1 (ADR1), ADR1-L1 and ADR1-L2—transduce signals that lead to SA accumulation and induction of downstream WRKYs modulated resistance. In rice, the CNL receptor, panicle blast 1 (Pb1), also appears to mediate resistance against rice blast in a mechanism involving interaction with WRKY45, a TF involved in induced resistance via SA signalling pathway. Some CNLs directly translocate or localize in the nucleus to activate defence, e.g., barley mildew A 10 (MLA10) and Arabidopsis RPS4 and RPS6. In the nucleus, MLA10 interacts with Hordeum vulgare (Hv) WRKY1/2, which are suppressors of basal defence, during incompatible interaction with powdery mildew fungus. A CNL designated as MLA1, also from barley, functions in Arabidopsis against Blumeria graminis f. sp. hordei (Bgh). The MLA1-triggered immunity, including host cell death response and disease resistance, is fully retained in Arabidopsis mutant plants that are simultaneously impaired in well-characterized defence-phytohormone pathways (ET, JA and SA). Similar to MLA1, co-acting Arabidopsis TNL pair, RPS4 and RRS1 (which encodes a WRKY DNA binding domain), confers resistance in cucumber, N. benthamiana, and tomato.

Another example supporting our understanding of the NLR nuclear activity is the interaction of N immune receptor with the TF SQUAMOSA PROMOTER BINDING PROTEIN-LIKE 6 (SPL6) in N. benthamiana. The N immune receptor is present in the nucleus, and confers resistance to tobacco mosaic virus (TMV) infection. N receptor associates with SPL6 at the sub-nuclear bodies only when the cognate effector, p50, is present in the cell. A genetic requirement for SPL6 was not only shown in N. benthamiana for N-mediated disease resistance using the yeast two-hybrid system, but also in A. thaliana for RPS4 immune receptor mediated defence against P. syringae pv.

tomato expressing AvrRps4 effector. Moreover, a number of RPS4-mediated defence responsive genes were differentially regulated upon AtSPL6 silencing, including some of the previously characterized defence responsive genes such as PAD4, PR1, ALD1, AIG1, NUDT6 and FMO1. Additional evidence has been shown in Arabidopsis RPW8 resistance protein, which encodes truncated CNL-like proteins conferring resistance to powdery mildews in N. tabacum and N. benthamiana as in Arabidopsis. RPW8 requires SA, EDS1, NPR1 and PAD4 to be effective. The functional role of RPW8 is typically similar to a TNL ADR1, a close homolog of N Requirement Gene 1 (NRG1), which functions in and beyond innate immunity. These findings present a unique opportunity to further understand how effector-activated immune receptors directly associate with TFs in the nucleus to activate immune responses. Overall, a resistance signalling framework appears to have emerged for plants in which certain specificity-determining (sensor) NLRs initiate the immune response and either auto-activate and contribute to defence or compliment with other signalling NLRs to contribute to defence by conveying or amplifying the signal.

## Phytohormones in Plant Defence Response to Pathogens and Insects

Plant defence against pathogen/herbivore attack involves many signal transduction pathways that are mediated by a network of phytohormones. Phytohormones also play a critical role in regulating plant growth and development. Three most reported plant defence response phytohormones against pathogens/insects include salicylic acid (SA), jasmonic acid (JA) and ethylene (ET). Salicylic acid, a benzoic acid derivative, is an extensively studied important phytohormone in the regulation of plant defence. In Arabidopsis, activation of the SA pathway has been shown to be important in both basal and R gene mediated biotrophic and hemibiotrophic pathogen defence. NDR1 and EDS1 act upstream of SA, while the downstream pathway is modulated by NONEXPRESSOR OF PR GENES 1 (NPR1), and WRKY45 in rice. NPR1 is a transcriptional co-activator of a large set of defence-related genes downstream of SA, and it can conditionally regulate PDF1.2 expression following treatment of plants with SA and MeJA. SA also contributes to the HR-associated resistance via mechanisms that interact with RBOHD, a catalyst in ROS generation and cell death. In tobacco, SA significantly increases in resistant plants infected with TMV. A similar response was observed in Ny-1-resistant potatoes after infection with Potato virus Y (PVY).

In response to insect attack, SA regulates plant defence signalling against aphids by modulating the activity of PAD4. Indeed, pad4 mutants, with compromised SA signalling, have increased susceptibility to Myzus persicae. Correspondingly, there is a correlation between pad4 susceptibility and a delay in aphid-induced senescence, indicating that SA defence pathways are compromised in pad4 mutants. Basal SA defences have also been shown to decrease M. euphorbiae longevity in tomato. Moreover, SA is necessary for Mi1.2-mediated resistance to potato aphids. SA is also a key derivative of SAR in plants. SAR is a 'whole-plant' broad-spectrum resistance response that occurs following an earlier localized exposure to a pathogen. It is well known that ETI can trigger SAR through both local and systemic synthesis of SA, resulting in transcriptional reprogramming of a battery of genes encoding PR proteins. The reports published so far point to different compounds as potential SAR signals. A change in amino acid homeostasis is one of the suggested components in SAR mediated by ETI. Moreover, amino acids have been reported to be precursors of a large array of plant secondary metabolites involved in defence, including signal SA, cell wall components and anthocyanins. Further evidence on the involvement of amino acid homeostasis in plant defence was reported in Arabidopsis agd2-like defence response protein 1 (ald1) mutants. Characterization of the Arabidopsis ald1 suggested that an amino acid–derived defence signal was generated upstream of SA synthesis. These findings reveal that plants likely employ amino acids and their derivatives to rapidly reprogram SA synthesis and cellular transcription in order to cope with pathogen invasion, even though it appears to be at the expense of growth and development.

SA also interacts with other phytohormones either synergistically or antagonistically. There is an obvious cross-talk between JA and SA signalling pathways in pepper to control thionin synthesis as part of the PR response and other defence pathways. Other synergistic examples include the treatment of N. benthamiana plants with JA or SA, which was shown to enhance systemic resistance to TMV; Ellis et al. have also shown that SA and JA-signalling pathways are required to accomplish the defence response necessary to avert pathogen attack. More recently, Arabidopsis mutants with constitutive SA responses were reported to require JA and ethylene signalling for SA mediated resistance. A dominant mutant named suppressor of SA insensitivity (ssi1), which has constitutive expression of PR genes and is resistant to P. syringae, was also shown to constitutively express PDF1.2 and accumulate elevated levels of SA. Although this finding may be intriguing, because SA does not normally induce PDF1.2 in wild-type plants, it suggests the existence of an intricate signalling network involving SA and JA. Another mutant named constitutive PR 5 (cpr5) was shown to have SA-mediated NPR1-independent resistance, which apparently required components of the JA and ET signal pathways. The pre-treatment of plants with JA followed by SA was also shown to remarkably enhance resistance more than otherwise. Moreover, plants impaired in the JA pathway fail to accumulate SA in the leaves or phloem and become highly susceptible to TMV. Conversely, impairing the SA pathway does not affect JA levels, although increased susceptibility is observed. During infection by the pathogen P. syringae pv. tomato (Pst) DC3000/AvrRpm1, JA as a systemic signal for SAR, increases significantly 6 hours after infection and returns to normal 11 hours after infection, which suggests that JA may be transiently required for SA accumulation. Further evidence indicates that SAR is compromised in JA-insensitive mutants, sgt1b/jai4, opr3 (JA-biosynthesis mutant) and jin1 (JA-response mutant). The JA-biosynthesis mutants dde2 and opr3 as well as the downstream signalling mutants coi1, jar1 and jin1, though intact in SAR, partially require JA biosynthesis for an effective resistance response. Thus, it is possible that JA probably modulates early components of the SA biosynthetic or signalling pathway. However, it seems likely that the synergistic mechanisms may require not only SA and JA, but also ethylene, considering that cpr5 phenotype is suppressed by the ethylene-insensitive (ein2) mutation.

The negative crosstalk between SA and JA/ET pathways is probably modulated by TGA1A-RELATED GENE (TGA) factors. TGA class of bZIP TFs are repressed by plant-specific glutaredoxins (e.g., ROXY19), which are in turn induced by SA. Co-expression of ROXY19 with OCTADECANOID-RESPONSIVE ARABIDOPSIS AP2/ERF-domain protein 59 (ORA59) and ETHYLENE INSENSITVE 3 (EIN3) complex suppresses ORA59 promoter activity. Moreover, SA negatively regulates ORA59 protein accumulation in 35S:ORA59-GFP overexpressing plants. ORA59 is a transcriptional regulator of JA/ET-induced defence genes and is activated by either JA or ET and suppressed by SA. More recently, TGA2, TGA5 and TGA6 were shown to activate the SA-suppression of ET-inducible defence by regulating ORA59 expression. This suggests that SA-suppresses JA/ET-inducible defence by interfering with ORA59 activity through regulation of ROXY-TGA interaction. Conversely, evidence of SA positive regulation of ET was proposed by Guan et al. These authors have shown that in Arabidopsis, SA modules ET by potentiating MITOGEN-ACTIVATED PROTEIN KINASE6 (MPK6) and MPK3, and involves two 1-aminocyclopropane-1-carboxylic acid synthase (ACS; ACS2 and ACS6) isoforms, which are downstream components o MPK signalling pathway. This finding adds another level of complexity to the phytohormones regulatory network and will probably require further elucidation on how this pathway differs from the ORA59 regulated pathway.

On the other hand, most ET dependent defenses are positively modulated by JA. The JASMONATE ZIM-DOMAIN (JAZ) protein, which directly binds EIN3/EIL1 and recruits HISTONE DEACETYLASE 6 (HDA6) to repress ET responsive transcription, is repressed in the presence of JA. Thus, accumulation of JA degrades JAZ and allows the binding of EIN3 to the ERF1 promoter resulting in the transcription of ERF1. EIN3 also directly activates the promoter of ORA59 that regulates JA/ET-activated defence pathway. Studies on microarray analysis of Arabidopsis plants infected with Alternaria brassicicola revealed that nearly half of the genes induced by ET are also induced by JA. This was substantiated by Lorenzo et al. who reported that JA and ET pathways indeed converge in the transcriptional activation of ERF1, which encodes a TF that regulates the expression of pathogen response genes. ERF TFs have been reported to exhibit different regulatory roles depending on the species. For instance, in wheat ERF gene TaPIEP1/TaPIE1, which belongs to the B3 subgroup within the ERF subfamily, confers enhanced resistance to the fungal pathogens, Bipolaris sorokiniana and R. cerealis, when overexpressed in transgenic wheat, whereas in cotton GhERF of group IX, which includes ORA59, confer resistance to Xanthomonas campestris pv. malvacearum. Because ERF1 integrates signals from the JA and ET defence signalling pathways, the constitutive expression of ERF family members activates the expression of several JA/ET-dependent defence genes and induces resistance against necrotrophic pathogens. For instance, expression of several PR genes which confer resistance against several necrotrophs (e.g., PR3 and PR5d and PDF1.2) is modulated by ERFs. These defence genes possess a GCC box in their promoters, which is a direct target for the action of ERFs.

Although ET has been shown to regulate plant defence responses against fungi and bacteria, ET is probably not essential in plant resistance against viruses. Recently, 1-aminocyclopropane-1-carboxylic acid (ACC) was shown to enhance TMVcg accumulation in treated plants, which increased susceptibility, suggesting that ET is required for viral infection.

Other phytohormones, such as ABA, gibberellins (GBs), auxins, brassinosteroids and cytokinins (CKs), have recently emerged as defence regulators. ABA, a sesquiterpene compound resulting from the cleavage of γ-carotene, regulates numerous developmental processes and adaptive stress responses in plants. ABA can positively regulate plant defence at the early stages of infection by mediating stomatal closure against invaders, or inducing callose deposition if the pathogen evades the first line of defence. If activated at later stages, ABA can suppress ROS induction and SA or JA signal transduction, thereby negating defences controlled by these two pathways.

Cytokinins promote cell division, and are known to play a role in the synthesis and maintenance of chlorophyll and chloroplast development and metabolism. CKs are also involved in the modulation of defence mechanisms, including the induction of resistance against viruses, but are known to suppress HR. Cytokinins can however act synergistically with SA signalling. CKs activate the transcriptional regulator ARABIDOPSIS RESPONSE REGULATOR 2 (ARR2), which positively modulates SA signalling by interacting with the SA-responsive factor TGA3. TGA3 induces the binding of ARR2 to the promoters of PR-1 and PR-2 to induce cytokinin-dependent gene transcription. Correspondingly, the npr1-1 or NahG mutants fail to modulate the induction of ARR2 when treated with CK, indicating that CK modulates signaling components downstream of SA. Moreover, increased transcription of genes involved in SA-biosynthesis and signalling (e.g., SID1, SID2, PR-1 and PR-5) is observed in ARR2 over-expressing mutants challenged with P. syringae pv. tomato (Pst DC3000). Thus, CKs synergistically interacts not only with the SA signaling pathway to boost SA dependent induction of plant defence genes but also modulates SA biosynthesis. Cytokinins have also been shown to enhance the production of two antimicrobial phytoalexins, scopoletin and capsidiol in tobacco plants challenged with P. syringae pv. tabaci (Pst) independent of SA signalling. Moreover, cytokinins induce the expression of cell wall invertase, a key sucrose cleaving enzyme required for carbohydrates supply through an apoplasmic pathway. Invertase is required for plant defence against pathogens, including Pst. The glucose target of rapamycin (TOR) signalling pathway involved in autophagy apparently modulates the transcriptional dynamics associated with cytokinin invertase induced defence pathway by providing the required energy, metabolites and the cell cycle machinery required for cytokinin signal transduction. The link between autophagy and cytokinin signalling was previously suggested, but the cytokinin-induced defence system in this interplay is probably a protective mechanism to maintain plant growth and proliferation despite pathogen challenge.

Brassinosteroids (BRs) are a class of polyhydroxysteroids that affect many cellular processes including elongation, proliferation, differentiation, membrane polarization and proton pumping. BRs are increasingly becoming important in plant defence against pathogens. The mechanism underlying BR signalling involves the direct binding of BRs such as BL and castasterone to the LRR-RLK (BRI1). This interaction is reported to unlock BRI1 from the negative regulator BKI1, followed by heterodimerization of BRI1 with a co-receptor BAK1 and phosphorylation of the BRI1-interacting signalling kinase (BSK1). Other events include the activation of the protein phosphatase BSU1. These biochemical changes inhibit the shaggy-like kinase BIN2, which culminates into the activation of the homologous TFs, BZR1 and BES1/BZR2. These TFs translocate to the nucleus, interact with BR-responsive promoters, and cause transcriptional changes that eventually lead to defence response. BRs have been demonstrated to enhance plant defence against pathogens. In potato, BRs have been shown to be effective against viral infection from the starting planting materials to the second tuber generation. Furthermore, application of BRs on tobacco plants decreases TMV viral load and restricts infection by other biotrophs. BAK1 is essential for plant basal immunity during compatible interactions with RNA viruses. The BAK1 mutants, bak1-4 and bak1-5, accumulate turnip crinkle virus (TCV), oilseed rape mosaic virus (ORMV) and TMV to higher levels compared to the WT plants. Thus, BAK1 could probably be a general regulator of plant defence against biotrophs and hemibiotrophs. BRs have also been reported to interact with other phytohormones, such as GA and auxins, but independent of SA.

Taken together, the intricate cross-talk among hormones to cooperate with other signals and to coordinate appropriate induction of defences against pathogens and insect pests depends on the pathogen type, physiological stage and environmental and probably circadian regulations.

## RNAi-mediated Plant Defence

RNA interference or silencing is one of the emergent crop improvement strategies that involve sequence-specific gene regulation by small non-coding RNAs, which mainly belong to two categories, i.e., small interfering RNA (siRNA) and microRNA (miRNA). Though these sRNAs differ in biogenesis, both regulate the target gene repression through ribonucleoprotein silencing complexes. Plant RNA silencing involves four basic steps, which include introduction of double-stranded RNA (dsRNA) into the cell, processing of dsRNA into 18–25-nt small RNA (sRNA), sRNA 2-O-methylation and sRNA incorporation into effector complexes that interact with target RNA or DNA. The formation of RNA-induced silencing complex (RISC) and its incorporation into the antisense strand of siRNAs, which interacts with Argonaute and other effector proteins, precedes the cleavage of the target mRNA. For sRNA to meet the target mRNA, it has to move from the point of initiation to the target. Thus, two main movement categories include cell-to-cell (short-range; symplastic movement through the plasmodesmata) and systemic (long-range; through the vascular phloem) movement. These mobile silencing strategies use sRNAs to target mRNA in a nucleotide sequence specific manner. By use of fluorescently labelled 21 and 24-nt siRNAs, Dunoyer et al. demonstrated the movement of siRNAs from cell to cell and over long distances. Such systematic movements enhance systemic silencing of viruses as reported in N. benthamiana. Similar systemic movements have been reported in the phloem sap of oilseed rape and pumpkin. Endogenous 21-nt miRNAs (miR399) were also reported to be mobile within the roots, and between shoots and roots of rapeseed and pumpkin. Thus, sRNAs can be targeted to most active plant tissues, with transcription activity, to achieve a desirable consequence.

Several RNAi strategies have shown success in plant improvement against biotic stresses. Arabidopsis miR393 was the first sRNA implicated in bacterial PTI, and enhanced miR393 accumulation was found during sRNA profiling in Arabidopsis challenged with Pst. The mechanism of miR393-induced resistance involves repression of auxin signalling by negatively regulating the F-box auxin receptors like transport inhibitor response 1 (TIR1). This process restricts Pst infection, and, indeed, plants overexpressing miR393 exhibit effective resistance against Pst.

RNAi in plant resistance to fungi has also shown promise. For instance, RNAi-mediated suppression of a rice gene OsSSI2 enhances resistance towards M. oryzae and X. oryzae. Moreover, RNAi suppression of OsFAD7 and OsFAD8, the two genes encoding for Ω-3 fatty acid desaturase, also enhances resistance against M. oryzae. RNAi targeting of lignin production pathway genes aimed at reducing lignin content has also been shown to enhance resistance against Sclerotinia sclerotiorum in soybean. Increased resistance to Blumeria graminis f. sp. tritici in wheat was also demonstrated through RNAi using 24 miRNAs. Nevertheless, the performance of these approaches under environmental conditions has often been unsatisfactory and environmental influences in expression of resistance often remain unpredictable.

In response to virus infection, several cases have shown successful crop improvement. For instance, resistance to African Cassava Mosaic Virus (CMV) was achieved in transgenic cassava plants producing dsRNA against PSTVd sequences. A similar strategy was successful in transgenic tomato resistance against Potato Spindle Tuber Viroid (PSTVd). RNAi targeting of the virus coat protein has also been successfully engineered into plants to induce resistance against viruses. For instance, transgenic tobacco plants expressing the CP gene of TMV are resistant to TMV.

In functional biology studies, virus-induced gene silencing (VIGS) has emerged to be one of the most powerful RNA-mediated post-transcriptional gene silencing (PTGS), not only in plant protection against viruses, but also for gene knockouts in functional genomic studies.

Although RNAi has the potential to contribute to increased crop productivity, by generating crops with improved resistance against pests and diseases, it would be even better if interaction between sRNAs and their targets is validated in several backgrounds. This would provide valuable insight into mechanisms of post-transcriptional gene regulation and multiple molecular pathways controlling plant stress responses. However, the danger of unintentional silencing of genes with regions of homology to the intended target, and target mutations leading to easier escape from miRNA-directed silencing are still ethical issues. Certain biosafety concerns on the use of RNAi transgenics, especially transcriptional gene silencing by chromatin modification is even a more sensitive and contentious issue, as it is rumoured to lead to hereditary changes associated with adverse effects. Thus, the underlying mechanisms associated with RNAi require further investigations using well-controlled experiments.

## Modern Approaches for Improving Biotic Stress Tolerance in Plants

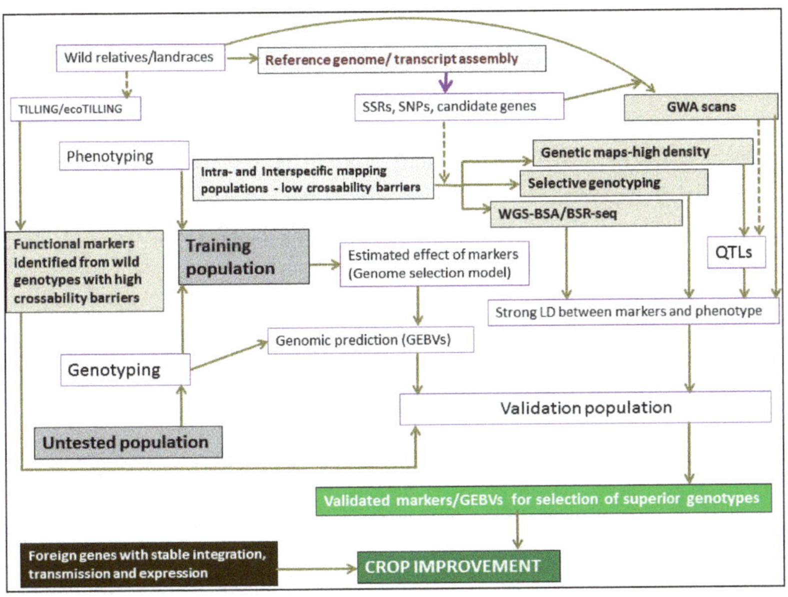


Conventional breeding methods still play an important role in the selection of new varieties. However, emerging tools in biotechnology are much needed to maximize the probability of success. One area of biotechnology, molecular marker assisted breeding (MAB), has already made significant impact in improving efficiency of conventional breeding. There are, however, major gaps in the improvement of traits controlled by a large number of small effects, epistatic QTLs displaying significant genotype × environment (G × E) interactions. Thus, accurate indirect selections based on genomic tools that have emerged over the last few decades are continuously being employed to improve the breeding efficiency for such traits. The advantage is that, to date, the genome sequences for more than 55 plant species have been produced and many more are being sequenced. The genome sequence information available enables the identification and development of genomewide markers. Availability of markers covering the whole genomic regions has already shown promise in the development of special populations, such as recombinant inbred lines (RILs), near isogenic lines (NILs), introgression lines (ILs) or chromosome segment substitution lines (CSSLs). Recently, heterogeneous inbred family (HIFs) and multi-parent advanced generation inter-cross (MAGIC) populations, which can serve the dual purpose of permanent mapping populations for precise QTL mapping and for direct or indirect use in variety development, have shown promise in plant breeding. Also, genomewide association (GWA) analysis has been successfully applied to rice, maize, barley, wheat, sesame and other plants. GWA has also been adapted to the "breeding by design" approach, often referred to as genome selection, which predicts the outcome of a set of crosses on the basis of molecular markers information.

Principle of genomic selection: Two steps are involved; developing a training population to provide phenotypic and genotypic data; effects are estimated for all molecular markers. The second step involves genotyping untested populations and selecting superior genotypes based on their expected phenotypes according to the estimates obtained from the marker effects on the training population (bottom).

Recently, a combination of different approaches has been used to develop new rice cultivars referred to as 'Green Super Rice', possessing resistance to multiple insects and diseases, high nutrient efficiency and drought resistance. If fully exploited, the integration of a similar approach with breeding by design or genome selection would help to design new plant types with not only a few selected major loci, but nearly all the functional loci of the genome controlling key desirable traits in commercial cultivars.

Expression studies also present a major area of interest for breeders. Among them, the NGS technologies have become the mainstay of studying complex traits, as direct sequencing of genomes and comparison with reference sequences is increasingly becoming more feasible. Re-sequencing has been performed for model species, e.g., Arabidopsis, to understand the whole genome sequence variation, and ultimately discover single nucleotide polymorphisms (SNPs). Similar re-sequencing efforts have been applied in rice, maize, soybean, grape and poplar. Combining re-sequencing with the recent developments in omic biology, including transcriptomics, proteomics, metabolomics, epigenetics and physiological and biochemical methods will remarkably provide novel possibilities to understand the biology of plants and consequently to precisely develop stress tolerant crop varieties.

The recent advent of genotyping by sequencing (GBS) approach that minimizes ascertainment biases and the need for prior genome sequence information associated with traditional techniques has also enabled single nucleotide polymorphism marker detection, exposition of QTLs and the discovery of candidate genes controlling stress tolerance. Thus, genome/transcript profiling when combined with genome variation analysis is a potential area which could prove useful for breeders in the near future. Another newly developed approach, which combines genetical genomics and bulk segregant analysis (BSA) to identify markers linked to genes, shows the possibility of coupling BSA to high throughput sequencing methods. Although there are shortcomings, including errors introduced during NGS procedures, this method has proven to be useful in identifying stress tolerance genomic regions in crop plants. A more recent modification that exploits the power of deep sequencing of target-enriched SNP markers to increase the efficiency of BSA analysis is called target-enriched TEX-QTL mapping. by combining a large F2 population size, deeply sequenced markers, and 10–20% bulk size, most QTLs can be identified within two generations. Although it does not currently detect very closely linked QTL, TEX-QTL method is potentially a useful development in plant breeding. It is envisaged that BSA, by genotyping pooled-segregant sequencing, is likely to increase the reliability and reduce the time required to map all QTL defining the trait of interest and to identify causative superior alleles that can subsequently be used for crop improvement by targeted genetic engineering.

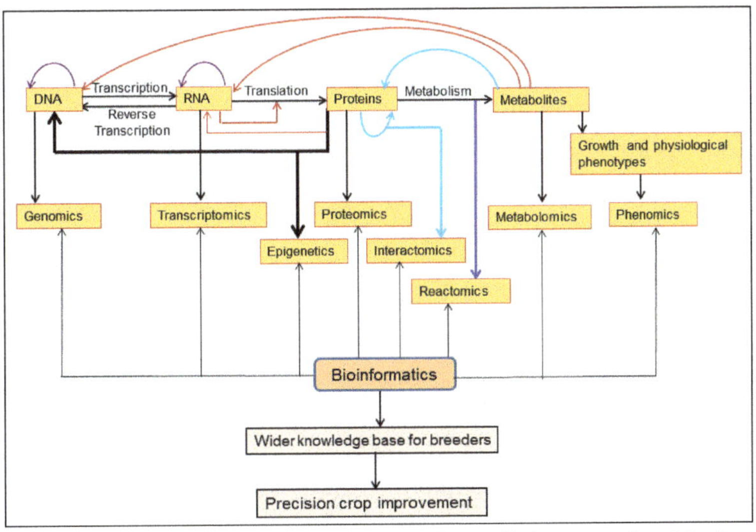


Supportive omic tools for increasing plant breeding efficiency against biotic stresses. Sky blue lines indicate interactions; largest bold black lines indicate epigenetic regulation; red lines indicate regulation; and blue line indicates metabolic reactions.

Desirable alleles are also being identified using functional genomic tools, including transformation, insertional mutagenesis, RNAi, the screening of either mutant or natural germplasm collections by means of targeting induced local lesions in genomes (TILLING) or ecotype TILLING (EcoTILLING) methodologies. These strategies enable plant scientists to predict gene functions and allow efficient prediction of the phenotype associated with a given gene, the so-called reverse genetics approach. The availability of a large volume of sequences generated through NGS technologies is significantly increasing the number and quality of candidates for TILLING and EcoTILLING studies. Thus, a number of crops have benefited from these technologies, including Arabidopsis, lotus, barley, maize, pea, melon and rice.

The use of improved recombinant DNA techniques to introduce new traits in early phases of cultivar selection is also currently gaining momentum in plant biology. Techniques such as oligonucleotide-directed mutagenesis (oDM) as well as those based on zinc finger nuclease (ZFN), transcription activator-like effector nuclease (TALeN) and clustered regularly interspaced short palindromic repeat (CRISPR)/CRISPR-associated protein 9 (Cas9) system are all capable of specifically modifying a given target sequence leading to genotypes not substantially different from those obtained through traditional mutagenesis. The practical use of these techniques in developing countries and the performance of the germplasm developed through them under environmental conditions is yet to be fully demonstrated.

# FUNGAL BIOTIC STRESSES IN PLANTS

Commonly fungal diseases cause more damage than diseases cause by other pathogens. Several fungi cause various diseases in different crop fields and plants. Sometime these fungi can cause similar symptoms and confused to one another. There are some preharvesting fungal diseases such as Aspergillus, Alternaria, anthracnose, Cercospora (frogeye) leaf spot, charcoal rot, Choanephora blight (wet rot), damping-off, root-rot, downy mildew, Fusarium stem rot, Fusarium wilt, gray leaf spot, gray mold, Phytophthora blight, powdery mildew, Southern blight, Verticillium wilt, and white mold that are grown in different crops.

Table: Fungal diseases and their causal agents attacking on chilli plant at different crop stages.

| Crop stage | Common fungal diseases | Causal agent |
|---|---|---|
| Vegetative stage | | |
| Seedling | Damping off, Phytophthora blight, Fusarium wilt | Pythium spp., Phytophthora sp., Fusarium oxysporum |
| After transplantation | Frog eye leaf spot, damping off, Fusarium wilt | Cercospora capsici, Pythium spp., Fusarium oxysporum |
| Root diseases | Phytophthora root rot, Verticillium wilt, Rhizoctonia root rot | Phytophthora capsici, Verticillium sp., Rhizoctonia solani |
| Reproductive stage | | |
| Flowering | Powdery mildew | Leveillula taurica |
| Fruiting | Aspergillus, anthracnose, powdery mildew, Phytophthora pod rot, black mold | Aspergillus spp., Colletotrichum sp., Leveillula taurica, Phytophthora capsici, Alternaria spp |

The vegetable and cereal crop are recorded about the entire region and areas of Pakistan. So there are no clear-cut boundaries between each region and intermingling of crop filed occurs. Pakistan has four major phytogeographical regions. Major floristic zones are summarized below.

Table: Major chilli-producing phytogeographical regions of Pakistan.

| Regions | Subregions | Major areas in Pakistan |
|---|---|---|
| Saharo-Sindian | | Lower Sindh, some areas of upper Sindh, Central and Southern Punjab, Southern Balochistan and KPK plains, Rawalpindi, Islamabad |

| Irano-Turanian | Western subregions<br>Eastern subregions or Central Asia | Waziristan and North Baluchistan Upper portion of northern areas, Baltistan, Chitral |
|---|---|---|
| Sino-Japanese | | Kashmir, KPK, Der, Swat, Muzaffarabad, Hazaras |
| Indian | | East and west of river Jhelum, Lahore |

## Major Fungal Diseases causing Loss of Crops

Commonly fungal diseases cause more damage than diseases caused by other pathogens. Several fungi caused various diseases in field crops. Some major fungal diseases which are responsible for heavy loss to the field crops are as follows:

### Alternaria Spp.

Genus Alternaria is considered as seed-borne fungi, widespread, and highly cause of decline that usually infects chili plants. The loss of fruit yield has been recorded to be up to 100% under certain environment conditions. The disease is characterized by fairly firm, superficial lesions which often occur on berries near the pedicel. Lesions are tan at first, become dark brown to black with age, and remain localized. Under humid conditions provided by cold transit, fluffy gray tufts of fungus often occur on rachis and pedicels, occasionally without causing visible lesions. Symptoms of Alternaria begin as water-soaked, gray lesion on either the side or blossom-end of the fruit. As the lesion progresses, they darken and become covered with spores. Internal necrosis and mycelial growth occur on the seeds, placenta, and pericarp but are not noticed until the harvesting. Infection can occur through the flowers or insect injury, mechanical damage, chilling injury, sunburn, or blossom-end rot.

### Anthracnose/Dieback/Fruit Rot (Colletotrichum Spp.)

Anthracnose or dieback and fruit rot are caused by Colletotrichum sp. The fruit-rot diseases the diseases of crops which belong to the tropical regions has been reported heavy damage. It is remarked as major harmful disease of vegetable crop. Anthracnose also caused by Colletotrichum spp. is a major problem of ripened fruit (Hence, it is also called as ripe fruit rot.), occurs worldwide wherever cereal and other crops are grown, is severely infected by anthracnose which may cause yield losses up to 50%. It is a seed-borne fungus. It produces dark spot, water-soaked lesions that rapidly expand. In some cases, the lesions are brown and then turn black from the formation of setae and sclerotia. Infections spread rapidly during periods of excess irrigation or rain on immature pods.

### Aspergillus Spp.

Several fungi may be present on fruit after the postharvest and storage due to the relative humidity is not controlled during the storage process. There are a number of Aspergillus sp. recorded in different crop-growing areas of Pakistan including Aspergillus flavus, A. niger, A. fumigates, etc. Several fruit and vegetables are reported to be contaminated with molds and their toxic metabolites, and Aspergillus flavus is the predominant mold on collected samples in several cases. Aflatoxins are chemically classified in secondary metabolite which is mostly produced by Aspergillus flavus, A. bombycis, A. nomius, A. parasiticus, and A. tamarii in chili.

## Damping Off (Pythium Spp.)

Damping off is a very common disease in nursery seedling beds showing irregular patches. Pythium spp. cause this disease which is basically a soil-borne fungus. This disease is more severe in soils with poor drainage.

## Powdery Mildew (Leveillula Taurica)

Powdery mildew is caused by Leveillula taurica (asexual stage of fungus). It usually occurs in warm climates (65–95°F), both dry and humid weather. It affects the lower leaf surface, where the lesions are covered with a white to gray powdery growth. Disease progresses from the older leaves to younger leaves, and defoliation is a prominent symptom which reduces the size and number of fruits.

## Fusarium Wilt (Fusarium Oxysporum)

Fusarium sp. is a soil-borne fungus and can survive for several years. It is usually grown in wet soil and high temperature. It causes wilting of the plant and upward and inward rolling of the leaves, and as a result leaves turn yellow and die. It causes discoloration of the vascular system of plant, particularly in the lower stem and roots.

## Cercospora Leaf Spot (Frog Eye) (Cercospora Spp.)

The leaf lesions due to this disease are circular about 1 cm in diameter with brown border and light gray centers. Lesions are also appearing on stem, petioles, and peduncles. It is a seed-borne fungus. It is also spread by water, wind, and leaf to leaf contact. Severely infected leaves turn yellow and drop from the plant. The fungus usually survives from one season to another on crop debris.

## Phytophthora Blight (Phytophthora Spp.)

This disease is also known as Phytophthora root rot. Fungus affected all parts of plant. It commonly creates problem when soils are excessively wet either from over irrigation, heavy rain, or both. The most common symptom is a stem or collar rot by sudden wilting without foliar yellowing.

## Common Recommended Cultural Control

Integration of chemicals and biotic agents along with resistance for managing plant disease has been considered as a novel approach. The studies of chemical and biological management are valuable to comprehend and understand biodiversity. In general, isolating, identifying, and determining structures of new metabolites are fundamental to reveal their chemical potential, a first step to use, conserve, and protect them. Several attempts to control root-rot fungi could be amended including disinfestations of the soil and planting material with fungicidal chemicals, crop rotation with non-hosts of the fungus, and use of resistant cultivars.

- Crop rotation.
- Discard the weeds from the crop.
- Disease-resistant variety should be used.
- Same family crops should not be planted in the same field more than once every 3 years.
- Excessive irrigation prior to seedling emergence should be avoided and, after establishment, water should not stand in the field for more than 12 hours.
- Before sowing, the seed should be dried by artificial (machine) method or sunlight method.

## Chemical Control

- Disinfect containers, seed trays, and propagators thoroughly. Dettol is good but tea-tree oil and camomile tea make good alternatives and are safe to use on plants.
- The strobilurin fungicides azoxystrobin (Quadris), trifloxystrobin (Flint), and pyraclostrobin (Cabrio) have been labeled for the control of anthracnose fungus in chili crop.
- Spray twice at 10–15 days interval with 2.5 g mancozeb or 1 g carbendazim/lit of water for the control of Cercospora leaf spot.
- Soil sterilization by drenching the soil 4” deep with formaldehyde diluted 50 times with water or with some other effective chemical soil sterilant.
- Soil drenching with 1% Bordeaux mixture or 3 g copper oxychloride like Blue Copper per liter of water at 12 and 20 days after sowing is also useful.
- Seed treatment with 3 g captan or thiram per kg seed.
- Seed treatment with 4 g Trichoderma viride formulation combined with 6 g metalaxyl is highly effective.

- Foliar spraying with captafol at 0.2% followed by copper oxychloride at 0.25% and carbendazim at 0.1% during the start of anthracnose disease.

## References


- Tree-crop-plant-stress-a-primer-on-abiotic-and-biotic-stressors: cid-inc.com, Retrieved 14 June, 2020
- Plants-and-biotic-stresses, plant-physiology-2, stress-physiology-23668: biologydiscussion.com, Retrieved 22 May, 2020
- Advances-in-plant-tolerance-to-biotic-stresses, plant-genomics: intechopen.com, Retrieved 06 August, 2020
- Fungal-biotic-stresses-in-plants-and-its-control-strategy, abiotic-and-biotic-stress-in-plants: intechopen.com, Retrieved 25 April, 2020

# Abiotic Stress in Plants

CHAPTER 2

Abiotic stress is defined as the negative impact of non-living factors on living organisms in a specific environment. The major cause of poor plant growth and reduced crop yields includes abiotic stresses, such as drought, salinity, low and high temperatures. This chapter has been carefully written to provide an easy understanding of abiotic stress in plants.

Abiotic stress is defined as the negative impact of non-living factors on living organisms in a specific environment. The stresses include drought, salinity, low or high temperatures, and other environmental extremes. Abiotic stresses, especially hypersalinity and drought, are the primary causes of crop loss worldwide. In contrast to plant resistance to biotic stresses, which mostly depends on monogenic traits, the genetically complex responses to abiotic stresses are multigenic and thus more difficult to identify, control, and manipulate. Abiotic stresses are one of the most important challenges facing crop production and can occur from a shortage of an essential resource, excess amounts of toxic substances or due to changes in environments.

Use of microbes may be a promising alternative strategy to overcome the limitations to crop production brought by abiotic stress and for better plant health and protection because they are environmental friendly, cost-effective and can also provide protection against biotic stresses. However, most of the work done on use of the microbes for abiotic stress management is confined mostly up to the controlled condition. There is need to test these microbes in field for their ability to manage abiotic stresses.

However, for successful use of Trichoderma spp. against biotic and abiotic stresses call for discovery or production of highly efficient strains. Most of the work done on use of microbes for abiotic stress management is confined mostly up to the controlled condition. There is need to test these microbes in field for their ability to manage abiotic stresses.

Abiotic stress tolerant plant varieties particularly in rice are proving to be boon for the farmers. However, only host gene is not enough under high stress severity or where crop is exposed to multiple stresses like flooding, drought and diseases. Similarly microbes alone may not be enough to provide proper protection against abiotic stresses. Therefore evoking crop–microbe interactions within and around the roots of abiotic stress tolerant varieties may further enhance tolerance of these varieties to not only multiple abiotic stresses but also biotic stresses. Therefore there is need to evaluate these microbes along with host tolerance under field condition to manage multiple stresses.

The abiotic stress conditions that most adversely affect crop yield are associated with water deficiency ion imbalance and temperature extremes. Plants respond to environmental stresses at multiple levels such as molecular, cellular, tissue, anatomical, morphological, and whole-plant physiological levels. These responses to one or more stresses also vary with species and genotype. The plant response to stress also depends on the duration and severity of the event, as well as the age and developmental stage of the plant when the stress is imposed. Among the various abiotic stress conditions, water deficiency is the most devastating factor. Salinity is a major constraint on crop plant productivity. More than 800 million hectares of land throughout the world are salt affected, which accounts for 6% of the world total land area. Effective management of the available water resources through agronomic practices plays a major role in sustainable production under drought.

During abiotic stress the biosynthesis and accumulation of different molecules thought to have protective functions in the cells is induced. These molecules are thought to mediate their protective function by their interaction with, or stabilizing of, different cellular components such as membrane elements or proteins/enzymes whose structure or function are sensitive and can be damaged as a result of the low temperature. For some of these protective molecules the biosynthetic pathways were described and genes encoding specific key enzymes in the pathways were identified and subjected to molecular manipulations to modulate biosynthesis. In cases where these protective molecules contribute to plant chilling tolerance, key biosynthetic or regulatory genes could serve as potential targets for biotechnological manipulations for improving cold tolerance for fresh produce during postharvest storage.

Glycinebetaine is an effective protectant against abiotic stress in plants including chilling and freezing. Much progress was made in recent years toward improving plant tolerance to abiotic stress by genetic engineering of glycinebetaine biosynthesis. Overproduction in tomato of glycinebetaine, otherwise not synthesized in tomato, resulted in plants more tolerant to chilling. This was obtained by over-expression of the codA gene, which encodes choline oxidase catalyzing the conversion of choline to glycinebetaine.

Heat shock pre-treatment in some plants has been found to induce induction of chilling tolerance and was demonstrated also to have a beneficial effect as a pre-storage treatment for some fruits or vegetables. Heat-shock proteins (HSPs) act to stabilize proteins in the cell under temperature stress conditions by their activity as molecular chaperones. The HSPs support protein folding, translocation, assembly, and degradation in the cells during optimal and normal growth conditions. Their function is also vital under stress conditions such as extreme temperatures.

Dehydrins, a group of late embryogenesis abundant (LEA) proteins, are characterized by the presence of a lysine-rich amino acid motif. LEA proteins are proposed to act as chaperones in the cells and function to protect proteins and membranes under stress conditions leading to dehydration. Induction of the level of LEA proteins was associated with improved tolerance to low temperatures. The over-expression of multiple dehydrin genes was demonstrated to enhance tolerance to freezing stress in Arabidopsis.

Polyamines (PAs) such as tetramine spermine (Spm), thetriamine spermidine (Spd), and diamine putrescine (Put) are considered components of plants' defense mechanisms against different types of abiotic stresses. Results from different studies demonstrated an increase in PA levels following postharvest exposure of fresh produce to chilling stress, mainly in chilling-injury-tolerant fruits. Manipulation of genes involved in the biosynthesis of PAs can be used to alter PA levels. Over-expression of the apple spermidine synthase gene in pear confers multiple abiotic stress tolerance by altering the amount of PA. No reports are available yet regarding the chilling tolerance effects by using this approach. Interestingly, over-expression of yeast spermidine synthase was recently shown to increase tomato fruit shelf life, probably by reducing postharvest senescence and decay.

It is not clear yet to what extent the mechanisms and genes found to be involved in tolerance against low-temperature stress during growth are also involved in tolerance to the stress imposed on sensitive, fresh produce when dark stored at low temperatures that induce chilling injuries. Both membrane structure characteristics and the antioxidative system seem to play a major role in both of these low-temperature stress conditions. Furthermore, manipulation of these biological processes may benefit other aspects such as plant development and the ability of crop plants to cope with different abiotic stress conditions such as drought, salinity, or high temperatures.

A plant's first line of defense against abiotic stress is in its roots. If the soil holding the plant is healthy and biologically diverse, the plant will have a higher chance of surviving stressful conditions.

The plant responses to stress are dependent on the tissue or organ affected by the stress. For example, transcriptional responses to stress are tissue or cell specific in roots and are quite different depending on the stress involved.

One of the primary responses to abiotic stress such as high salinity is the disruption of the $Na^+/K^+$ ratio in the cytoplasm of the plant cell. High concentrations of $Na^+$, for example, can decrease the capacity for the plant to take up water and also alter enzyme and transporter functions. Evolved adaptations to efficiently restore cellular ion homeostasis has led to a wide variety of stress tolerant plants.

Facilitation, or the positive interactions between different species of plants, is an intricate web of association in a natural environment. It is how plants work together. In areas of high stress, the level of facilitation is especially high as well. This could possibly be because the plants need a stronger network to survive in a harsher environment, so their interactions between species, such as cross-pollination or mutualistic actions, become more common to cope with the severity of their habitat.

Plants also adapt very differently from one another, even from a plant living in the same area. When a group of different plant species was prompted by a variety of different stress signals, such as drought or cold, each plant responded uniquely. Hardly any of the responses were similar, even though the plants had become accustomed to exactly the same home environment.

Rice (*Oryza sativa*) is a classic example. Rice is a staple food throughout the world, especially in China and India. Rice plants experience different types of abiotic stresses, like drought and high salinity. These stress conditions have a negative impact on rice production. Genetic diversity has been studied among several rice varieties with different genotypes using molecular markers.

Serpentine soils (media with low concentrations of nutrients and high concentrations of heavy metals) can be a source of abiotic stress. Initially, the absorption of toxic metal ions is limited by cell membrane exclusion. Ions that are absorbed into tissues are sequestered in cell vacuoles. This sequestration mechanism is facilitated by proteins on the vacuole membrane.

Chemical priming has been proposed to increase tolerance to abiotic stresses in crop plants. In this method, which is analogous to vaccination, stress-inducing chemical agents are introduced to the plant in brief doses so that the plant begins preparing defense mechanisms. Thus, when the abiotic stress occurs, the plant has already prepared defense mechanisms that can be activated faster and increase tolerance.

## Salt Stress in Plants

Soil salinization, the accumulation of water-soluble salts to levels that negatively impact plant production, is a global phenomenon affecting approximately 831 million hectares of land. More specifically, the phenomenon threatens 19.5% of the world's irrigated agricultural land and 2.1% of the world's non-irrigated (dryland) agricultural lands. High soil salinity content can be harmful to plants because water-soluble salts can alter osmotic potential gradients and consequently inhibit many cellular functions. For example, high soil salinity content can inhibit the process of photosynthesis by limiting a plant's water uptake; high levels of water-soluble salts in the soil can decrease the osmotic potential of the soil and consequently decrease the difference in water potential between the soil and the plant's roots, thereby limiting electron flow from $H_2O$ to P680 in Photosystem II's reaction center.

Over generations, many plants have mutated and built different mechanisms to counter salinity effects. A good combatant of salinity in plants is the hormone ethylene. Ethylene is known for regulating plant growth and development and dealing with stress conditions. Many central membrane proteins in plants, such as ETO2, ERS1 and EIN2, are used for ethylene signaling in many plant growth processes. Mutations in these proteins can lead to heightened salt sensitivity and can limit plant growth. The effects of salinity has been studied on Arabidopsis plants that have mutated ERS1, ERS2, ETR1, ETR2 and EIN4 proteins. These proteins are used for ethylene signaling again certain stress conditions, such as salt and the ethylene precursor ACC is used to suppress any sensitivity to the salt stress.

## Phosphate Starvation in Plants

Phosphorus (P) is an essential macronutrient required for plant growth and development, but most of the world's soil is limited in this important plant nutrient. Plants can utilize P mainly in the form if soluble inorganic phosphate (Pi) but are subjected to abiotic stress of P-limitation when there is not sufficient soluble PO4 available in the soil. Phosphorus forms insoluble complexes with Ca and Mg in alkaline soils and Al and Fe in acidic soils that makes it unavailable for plant roots. When there is limited bioavailable P in the soil, plants show extensive abiotic stress phenotype such as short primary roots and more lateral roots and root hairs to make more surface available for Pi absorption, exudation of organic acids and phosphatase to release Pi from complex P containing molecules and make it available for growing plants organs. It has been shown that PHR1, a MYB related transcription factor is a master regulator of P-starvation response in plants. PHR1 also has been shown to regulate extensive remodeling of lipids and metabolites during phosphorus limitation stress.

## Drought Stress

Drought stress is one of the main causes of crop loss within the agricultural world. This is due to water's necessity in so many fundamental processes in plant growth. It has become especially important in recent years to find a way to combat drought stress. A decrease in precipitation and subsequent increase in drought are extremely likely in the future due to an increase in global warming. Plants have come up with many mechanisms and adaptations to try and deal with drought stress. One of the leading ways that plants combat drought stress is by closing their stomates. A key hormone regulating stomatal opening and closing is abscisic acid (ABA). Synthesis of ABA causes the ABA to bind to receptors. This binding then effects the opening of ion channels thereby decreasing turgor pressure in the stomata and causing them to close. Recent studies, by Gonzalez-Villagra, et al., showed how ABA levels increased in drought-stressed plants. They showed that when plants were placed in a stressful situation they produced more ABA to try and conserve any water they had in their leaves. Another extremely important factor in dealing with drought stress and regulating the uptake and export of water is aquaporins (AQPs). AQPs are integral membrane proteins that make up channels. These channel's main job is the transport of water and other necessary solutes. AQPs are both transcriptionally and post transcriptionally regulated by many different factors such as ABA, GA3, pH and $Ca^{2+}$ and the specific levels of AQPs in certain parts of the plant, such as roots or leaves, helps to draw as much water into the plant as possible. By understanding both the mechanism of AQPs and the hormone ABA, scientists will be better able to produce drought resistant plants in the future.

One interesting thing that has been found in plants that are consistently exposed to drought, is their ability to form a sort of "memory". In a study by Tombesi et al., they found plants who had previously been exposed to drought were able to come up with a sort of strategy to minimize water loss and decrease water use. They found that plants who were exposed to drought conditions actually changed they way the regulated their stomates and what they called "hydraulic safety margin" so as to decrease the venerability of the plant. By changing the regulation of stomata and subsequently the transpiration, plants were able to function better in situations where the availability of water decreased.

Examples:

Abiotic stress comes in many forms. The most common of the stressors are the easiest for people to identify, but there are many other, less recognizable abiotic stress factors which affect environments constantly.

The most basic stressors include:

- High winds.
- Extreme temperatures.
- Drought.
- Flood.
- Other natural disasters, such as tornadoes and wildfires.
- Cold.
- Heat.

Lesser-known stressors generally occur on a smaller scale. They include: poor edaphic conditions like rock content and pH levels, high radiation, compaction, contamination, and other, highly specific conditions like rapid rehydration during seed germination.

## ABIOTIC STRESS RESPONSES IN PLANTS

Plants are often subjected to unfavorable environmental conditions – abiotic factors, causing abiotic stresses that play a major role in determining productivity of crop yields but also the differential distribution of the plants species across different types of environment. Some examples of abiotic stresses that a plant may face include decreased water availability, extreme temperatures (heating or freezing), decreased availability of soil nutrients and excess of toxic ions, excess of light and increased hardness of drying soil that hamper roots growth. The ability of plants to adapt and acclimate to different environments is directly or indirectly related with the plasticity and resilience of photosynthesis, in combination with other processes, determining plant growth and development, namely reproduction. A remarkable feature of plant adaptation to abiotic stresses is the activation of multiple responses involving complex gene interactions and crosstalk with many molecular pathways.

Abiotic stresses elicit complex cellular responses that have been elucidated by progresses made in exploring and understanding plant abiotic responses at the whole-plant, physiological, biochemical, cellular and molecular levels. One of the biggest challenges to modern sustainable agriculture development is to obtain new knowledge that should allow breeding and engineering plants with new and desired agronomical traits. The creation of stress-tolerant crop either by genetic engineering or through conventional breeding covered almost all aspects of plant science, and is pursued by both public and private sector researchers.

During the last decade, our research groups have focused their research on elucidating the different components and molecular players underlying abiotic stress responses of a broad range of species both model and crops plant. Several attempts to engineer those species with improved abiotic stress traits (drought and salinity) were made and the response of genetically engineered plants was deeply studied after establishment of adequate physiological methods. Now, we are moving efforts to expand our knowledge on plants response to abiotic stresses using holistic System Biology approaches, taking advantage of available high throughput tools such as transcriptomics, proteomics and metabolomics.

## Physiological Mechanisms Underlying Abiotic Stress Responses

Stress is a concept imported from physics. It was introduced in the theory of elasticity as the amount of force for a given unit area. In a biological context, stress is usually defined as an external factor that exerts a disadvantageous influence on the plant. Alternatively, stress could be defined as a significant deviation of the optimal condition of life.

## Physiological Responses to Early Abiotic Stress: Functional Decline in the Alarm Phase — The Stress Reaction

Three main phases may be considered on plant stress events and responses: (i) the phase of alarm; (ii) the phase of resistance; and (iii) the phase of exhaustion. Lichtenthaler added a fourth phase, the regeneration phase, which occurs only when the stressor is removed before damage being too severe, allowing partial or full regeneration of the physiological functions. The alarm phase starts with the so-called stress reaction, characterized by functional declines due to the stressor factor, offset by restitution counter reactions, in the transition to the phase of resistance. Stressors rarely act separately and individually on a plant. Generally, several stress factors act simultaneously, such as the frequently combined, at sunny, warm and dry summer periods, heat, water and high-light stress.

Sensing is the very first event experienced by a plant when one or more environmental factors (biotic or abiotic) depart from their optimum. Stress sensing is a complex issue and there is not a single sensing mechanism common to all stresses. For instance, some stresses directly affect the underground parts of plant bodies (e.g. drought, flooding) whereas other stresses (e.g., photoinbition) affect directly the aboveground structures of plant bodies. It is, thereby, expected that different sensing mechanisms will be involved. The most common model of sensing external stimuli is that of a chemical ligand binding to a specific receptor. This model, however, is suitable only for chemical stresses (e.g., heavy metal stress, nutrient depletion stress), not for physical stresses: primary sensing of temperature stress (heat stress or chilling/freezing) do not involve any chemical ligand. The same applies to radiation stress, although in this case an analogy between "ligand – receptor" and "photon – receptor" could be made. Even when molecules are involved, the universal character of the ligand - receptor model is debatable. In fact, in what concerns the rooting system, it is unclear if cells can sense the water concentration in the soil. In contrast, experimental evidences point to the possibility of sensing cell water homeostasis. The isolation of a transmembrane hybride-type histidine kinase from Arabidopsis thaliana provides experimental evidence for osmosensors in higher plants. Also sugars generated by photosynthesis and carbon metabolism in source and sink tissues play an important role in sensing and signaling, modulating growth, development, and stress responses.

Following sensing, one or more signaling and signaling transduction cascades are activated, preparing restitution counter reactions which will lead to the phase of resistance to stress. Meanwhile, functional declines are generally observed, including the photosynthetic performance, transport or accumulation of metabolites and uptake and translocation of ions. If these declines are not counteracted, acute damage and death may occur. The importance of restitution counter reactions is highlighted in experiments where different rates of stress imposition are compared: a more pronounced decline of physiological functions (photosynthesis, photosynthetic capacity and electron transport rate) was observed when higher plants were rapidly dehydrated than when the rate of water loss was slower. In desiccation resistant bryophytes there is a threshold of water loss rate behind which no physiological restoration is observed. Increased damage with more rapidly imposed stress is due, at least in part, to increased production of active oxygen species (AOS). Significant differences in the physiological behavior between the phase of alarm and the phase of resistance were highlighted by Marques da Silva and Arrabaça in, who found in the C4 grass Setaria sphacelata a decrease on the activity of the enzyme phosphoenolpyruvate carboxylase after several days of water stress, in sharp contrast with the several-fold increase of its activity observed after a short period of acute stress.

## Common and Distinctive Features of Salinity, Cold and Drought Stress

Salinity, cold and drought stress are all osmotic stresses: They cause a primary loss of cell water, and, therefore, a decrease of cell osmotic potential. However, the elicitor of cell water loss differs between stresses: (i) salinity stress decreases cell water content due to the decrease of external water potential, caused by the increased ion concentration (mainly $Na^+$ and $Cl^-$), turning more difficult water uptake by roots and water translocation to metabolically active cells; (ii) cold stress decreases cell water content due to the so-called physiological drought, i.e., the inability to transport the water available at the soil to the living cells, mainly the ones of the leaf mesophyll; (iii) the decrease of the cell water content under drought stress is due to water shortage in soil and in the atmosphere. Anyway, dehydration triggers the biosynthesis of the phytohormone abscisic acid (ABA) and it has been known for a long time that a significant set of genes, induced by drought, salt, and cold stresses, are also activated by ABA.

As a consequence of water loss and decreased cell volume, cell sap solute concentrations increase and thereby cell osmotic potential decreases. As cell turgor also decreases, an early effect common to these stresses is a sharp decrease in leaf expansion rate and overall plant growth rate. Furthermore, an additional active decrease of the cell sap osmotic potential is observed, as an attempt to keep cell hydration. In fact, at the metabolic level, a common feature to these three stresses is the osmotic adjustment by synthesis of low-molecular weight osmolytes (carbohydrates, betain and proline) that can counteract cellular dehydration and turgor loss. On the other hand, differences between these stresses do also exist. While drought stress is mainly osmotic, ion toxicity, namely $Na^+$, is a distinctive feature of salinity stress. Cold stress, behinds physiological drought, has an impact on the rate of most biochemical reactions, including photosynthetic carbon metabolism reactions, as enzyme activities are extremely temperature-dependent. Also water stress and salinity stress decrease photosynthesis, which create conditions to increased photoinhibition, particularly under high irradiances.

## Plant Bioenergetics as a Core to Stress Sensor

Despite the different physiological responses to early abiotic stress, a common point observed is the changes in the plant bioenergetic status. Such changes may involve a decrease in the energy production and an increase in energy demand to overcome the stress. The bioenergetics status is often considered as the chemical energy provided by adenylate energy charge (AEC), as defined in, for which plants are mainly dependent on photosynthesis.

The effect of abiotic stresses on photosynthesis can be perceptible: (i) within the photochemical reactions in the tylakoid membrane; (ii) in the carbon reduction cycle in the stroma; (iii) in the carbohydrate use in the cytosol and; (iv) on the $CO_2$ supply to the chloroplast dependent of stomata, mesophyll and chloroplast conductance. ATP and NADPH resulting from photochemical reactions are used in all others processes except $CO_2$ supply to the chloroplast in $C_3$ plants, so any limitation in photosynthesis such as those imposed by drought, can alter the plant bioenergetics status.

When the ATP and NADPH production by photochemical processes exceed the capacity for utilization in $CO_2$ fixation, plants can use several processes to dissipate energy and avoid or minimised photoinhibition. These processes include alternative electron sinks dependent of $O_2$ such as the oxygenase reaction catalised by ribulose-1,5-bisphosphate carboxylase/oxigenase which initiates photorespiration. The light-dependent $O_2$ uptake by photorespiration not only use ATP and reducing power from photosynthetic electron transport system but also cause a loss of the $CO_2$ fixed by Calvin cycle. Even in plants under no photoinhibitory conditions, photorespiration occur due to the capacity of Rubisco to catalise the carboxylation and oxygenation of ribulose-1,5-bisphosphate, depending on the $CO_2/O_2$ ratio. At 25 °C, photorespiration increases the cost of carbon (C) fixation to 4.75 ATP and 3.5 NADPH per C fixed under atmospheric $CO_2$ and $O_2$ Concentrations, which compares to 3 ATP and 2 NADPH per C fixed under no photorespiration conditions, e. g. only 2% $O_2$ instead of atmospheric 21% $O_2$. In plants submitted to drought, a reduction of photosynthesis and photorespiration is observed as a result of the lower $CO_2$ and $O_2$ availability in the chloroplast. However, in this situation, the photorespiratory pathway is less decreased than photosynthesis, as firstly suggested Lawlor and co-workers. In fact, despite the much higher affinity of Rubisco for $CO_2$ than $O_2$, the $CO_2$ concentration is almost at the sub saturating level in $C_3$ plants. Thus any decrease in stomatal conductance or in the gases solubility limits the carboxylase activity while the oxigenase activity is unaffected or less affected. In $C_4$ plants, the higher $CO_2$ concentration at the Rubisco level allows a lower decrease in the photosynthesis/photorespiration ratio under water deficit than the one observed in $C_3$ plants, despite the $C_4$ pathway having per se specific energy costs. The less efficient light use for $CO_2$ fixation caused by photorespiration lowers the quantum yields of photosynthesis in $C_3$ plants under drought or high temperature but this was not observed in $C_4$ plants. Since photorespiration is the major cause of a lower bioenergetic balance in photosynthetic tissues of $C_3$ plants, increasing plant growth by overcoming the limitation of photosynthesis imposed by Rubisco is still an important target of research and plant improvement.

In $C_3$ and $C_4$ plants under water deficit, the photosynthetic rate decreases with the leaf relative water content and water potential. This decrease is frequently correlated to the impairment of photochemical processes in $C_3$ plants, including inhibition of ATP synthesis. It is still unclear if photosynthesis is primarily limited by water deficit through the restriction of $CO_2$ supply to metabolism (stomatal limitation) or by the impairment of other processes which decrease the potential rate of photosynthesis (non-stomatal limitation). Nevertheless research efforts on these subjects are relevant to improve plants responses to stress.

Biochemical modeling of leaf photosynthesis in $C_3$ and $C_4$ plants can provide useful insights into the evaluation of stomatal and non-stomatal limitations of photosynthesis, as previously shown in drought stressed Medicago truncatula plants and Paspalum dilatatum plants under water deficit, elevated $CO_2$ and dark chilling. Photosynthesis light curves allow the determination of the relative contribution of respiration, photosynthesis and photorespiration to the light energy dissipation. Additionally, they are an expeditious method to screen plants with improved resistance to water deficit, as also shown with M. truncatula transgenic lines.

The role of plant mitochondria in the bioenergetic balance is complex and involves cytocrome c oxidase but also several other processes such as alternative dehydrogenases and alternative oxidase that are independent of the adenylate control. An increase in leaf respiratory energy demand to overcome the drought stress via respiration was referred in leaves in few studies. More often, in drought plants, no change or a decrease in respiration is observed in leaves but the variations were always minor comparing to photosynthesis, despite the interdependence of the two processes through photorespiration. However, at the whole-plant level, the contribution of respiration to the plant bioenergetics status is relevant because respiration can account for a release of 30-70% of the C fixed daily in well-watered plants, whereas in drought plants the proportion of C lost increases, mainly due to the decrease observed on photosynthesis.

## Stress Interaction: Photoinhibition

Photoinhibition, the decrease of photosynthesis and photosynthetic capacity due to exposure to excess photosynthetically active radiation, is dependent not only on the radiation level but also on the level of metabolic activity.

Thereby, all stresses that decreased energy demand increased photoinhibition. In fact, photoinhibition occurs when the demand from the carbon reduction cycle for ATP and, mainly, reductive power is decreased and, thereby, not enough $NADP^+$ is available to act as the terminal electron acceptor of the linear photosynthetic electron transport chain. In these circumstances, the photosynthetic electron transport chain becomes over-reduced and AOS such as hydroxyl radicals, the superoxide anion and hydrogen peroxide are formed, causing oxidative damage to the components of the photochemical apparatus. It is well established that the main target of oxidative damage is the D1 protein and that photoinhibition occurs when the accumulation of photooxidized D1 surpasses its de novo synthesis. Plants developed several mechanisms to cope with high irradiance and avoid photoinhibition. These range from the anatomical to the molecular level. Paraheliotopic leaf movements or leaf nastic growth, allowing the vertical orientation of leaves, optimizes the leaf to irradiation angle in order to decrease energy load and prevent photoinhibition. Leaf chloroplast movements, to minimize exposition to high irradiation or to fulfill auto-shading, represents another example of strategies to avoid or minimize photoinhibition.

At the molecular level, non-photochemical quenching of chlorophyll fluorescence regulates energy dissipation at the primary photosynthetic reactions and therefore constitutes the first protection line against photodamage. This dissipative pathway is controlled by the thylakoid lumen pH and the xanthophyll cycle which increases the dissipation of excitation energy by inducing an enzymatic conversion of the carotenoid violaxanthin into antheraxanthin and zeaxanthin. Additionally, a second line of defense is provided by alternative electron cycling such as photorespiration. When photooxidation cannot be avoided, damage in the photosynthetic apparatus occurs, especially in PSII, where the reaction center D1/D2 heterodimer is the main site to be affected, mainly D1 while D2 is affected in a lesser extent. The repair of damaged components is then activated, as D1 has a high turnover rate. However, if the rate of repair fails to keep pace with the rate of damage, photosynthesis is decreased and photoinhibition occurs. Nuclear-encoded early-light inducible proteins (ELIPs) may play a relevant role in the protection mechanism.

## Stress and Plant Life-cycle: The Case of Drought Stress

It is well known that drought stress at the early stages of plant life, shortly after germination, may have devastating impacts as both the root system is not yet fully established, in one hand, and stomatal control is not yet fine tuned. However, drought stress at this early life stage did not attract much research attention, because it is easily overcome by farmers through an accurate choice of seedling dates. Drought stress at later phenological stages received most attention, particularly the comparison between drought effects on the vegetative phases and in the reproductive phases over grain production. It is now well established that the effects of stress may vary significantly with the phenological stage of plants. Reproductive stages are generally more sensible to stress than vegetative ones, but differences can also be made between different phases of the reproductive stage. Mouhouche et al. found in Phaseolus vulgaris that periods of flowering were more sensitive than pod elongation and grain filling phases. Casanovas et al. reported a decrease of both leaf physiology and grain yield in maize subjected to drought during flowering. Boonjung and Fukai reported that when drought occurred during vegetative stages, it had only a small effect on subsequent development and grain yield. The effect of water stress on yield was most severe when drought occurred during panicle development.

Grapevine provides an interesting example of the complexity of the relationships between drought stress and plant phenology. Traditionally, grapevine is a non-irrigated crop that occupies extensive areas in dry lands and semi-arid regions. Recently, in the Mediterranean region, irrigation was introduced to increase the low land yield. However, wine quality is strongly dependent on the organoleptic characteristics of grapes which, in turn, particularly in what concerns soluble sugar contents, are dependent on moderate drought stress during berry expansion (i.e. in the phases from fruit set to veraison). The irrigation strategy must therefore maximize the vineyard production without decreasing berry quality, an objective suitable for deficit irrigation programs (DRI).

Furthermore, a deep understanding of plant carbon assimilation and partitioning mechanisms under different water regimes will be required in the frame of precision agriculture, as, in fact, these mechanisms play a key role in the fine tuning of the balance between berry yield and quality. Hopefully, this will lead to the adoption of criteria for irrigation scheduling based on vine physiology.

## Gene Expression and Regulation under Abiotic Stress

### Complexity of Gene Expression and Regulation

Plants have evolved intricate mechanisms at multiple levels that increase tolerance in order to adapt to adverse conditions and to an ever sessile living. Plant growth and productivity are affected to a great extent by environmental stresses such as drought, high salinity, and low temperature. Expression of a variety of genes is induced by these stresses in various plants. The products of these genes impact not only stress tolerance but also in stress response.

Genes induced during stress conditions function not only in protecting cells from stress by producing important metabolic proteins, but also in regulating genes for signal transduction in the stress response. The first group includes proteins that probably function in stress tolerance, such as chaperones or late embryogenesis abundant (LEA) proteins. The second group contains protein factors involved in further regulation of signal transduction and gene expression that probably function in stress response. In some cases networks and cascades of expression are activated in response to a stress condition. The regulation of the expression of these networks is being studied during the last decades. The use of microarray approaches, and more recently, of the Next Generation Sequencing (NGS) methodologies have unveiled new regulatory mechanisms that complicate the understanding and most of all the possibilities to modulate and control these processes in view of improving plant responses and productivity.

The regulation of plant genes can be observed at three levels: transcriptional; post-transcriptional and post-translational. In each level, actions depend on specific molecular elements as well as molecular networks and cascades.

The transcriptional regulation involves the interplay of three major elements: chromatin and its modification and remodeling; cis-regulatory elements which are often binding sites, such as enhancers and promoters, located upstream and downstream the coding region; and trans-regulatory elements, usually transcription factors. Chromatin modification and remodeling involved in plant abiotic stress response have been observed in numerous situations. The sensitization of stress responsiveness is called priming. Priming boosts the plant's defensive capacity and brings it into an alarmed state of defense. Recently, priming was correlated with chromatin modification of promoter region of WRKY transcription factors. The involvement of epigenetic mechanisms in the response to environmental cues and to different types of abiotic stresses has been documented. Recent reports have shown that different environmental stresses lead to altered methylation status of DNA as well as modifications of nucleosomal histones.

Promoters are regulatory regions of DNA located upstream of genes that bind transcription factor IID (TFIID) and allow the subsequent coordination of components of the transcription initiation complex, facilitating recruitment of RNA polymerase II and initiation of transcription.

Members of dehydration-responsive element-binding (DREB) or C-repeat binding factor (CBF), MYB, basic-leucine zipper (bZIP), and zinc-finger families have been well characterized with roles in the regulation of plant defense and stress responses. Most of these transcription factors (TFs) regulate their target gene expression through binding to the cognate cis-elements in the promoters of the stress-related genes. More recently the

WRKY transcription factors are becoming one of the best-characterized classes of plant transcription factors. Several WRKY proteins were shown to be involved in plant drought and salinity stress responses. For example, overexpression of the Oryza sativa WRKY11 under the control of Heat Shock Protein 101 (HSP101) promoter led to enhanced drought tolerance. Similarly, the altered salt and drought tolerance of 35S:OsWRK45 and 35S:OsWRK72 Arabidopsis plants may be attributed to induction of ABA/stress-related genes.

NAC (N-acetylcysteine) proteins are plant-specific TFs which have been shown to function in relation to plant development and also for abiotic and biotic stress responses. The cDNA encoding a NAC protein was first reported as the RESPONSIVE TO DEHYDRATION 26 (RD26) gene in Arabidopsis. For example OsNAC6 expression is induced by cold, drought, high salinity, and ABA. OsNAC6 showed high sequence similarity to the Arabidopsis stress-responsive NAC proteins ANAC019, ANAC055, and ANAC072 (RD26). It seems that abiotic stress-responsive NAC-type transcription factors, especially the SNAC group genes, have important roles for the control of tolerance against environmental stresses such as drought.

Post-transcriptional regulation is a second level of gene expression modulation which is represented by four groups of processes: Pre-messenger (mRNA) processing (capping, splicing, and polyadenylation), mRNA nucleocytoplasmic trafficking, mRNA turn-over and stability, and mRNA translation.

Alternative splicing is widely known to regulate gene expression in plants subjected to low and high temperatures. For example, it was shown that STABILIZED1 (STA1), a gene coding for a nuclear pre-mRNA splicing factor is important under cold stress conditions in A. thaliana. Alternative splicing has been reported upon water deficit as well.

Since the early 2000's, several reports have associated small RNAs to abiotic stress responses, showing that post-transcriptional regulation of gene expression plays an important role in these phenomena. Small RNAs (20 to 25 nt) are processed from non-coding double-stranded RNA precursors by RNAses of the DICER-LIKE (DCL) family and mediate a series of gene silencing mechanisms. One of these mechanisms cleaves mRNAs or prevents their translation through the mediation of 21 nt microRNAs. The discovery that stress can regulate microRNA (miRNA) levels, coupled with the identification of stress-associated genes as miRNA targets provided clues about the role of miRNAs in stress responses. Functional analyses have demonstrated that several plant miRNAs play vital roles in plant resistance to abiotic stresses.

Messenger RNA translation is dependent on mRNA cytoplasmic cycling namely compartmentalization in P bodies and association to ribosomes. The amount of mRNAs in polysomes is generally reduced during exposure to dehydration or anoxia, while stress-induced mRNAs significantly increase in polysome association. In chloroplasts, RNA binding proteins and several nucleases have been described to adjust the relative half-life of their mRNAs in response to environmental cues, particularly light conditions.

At the post-translational level phosphorylation, sumoylation and ubiquitination of proteins are processes that play major roles in the modulation of plant response to abiotic stress. Phosphorylation and de-phosphorylation play major roles in the responses to abiotic stress. Several signal transduction cascades formed by mitogen-activated protein kinases (MAPKs) and SNF-1-related protein kinases (SnRKs) are activated upon water deprivation and osmotic stress through the phosphorylation of specific residues. Among these, SnRK2 proteins have been shown to be involved in ABA-dependent responses to water deficit, like stomata closure.

The up-regulation of the XERICO gene, encoding a H2-type zinc-finger E3 ubiquitin ligase, results in increased drought tolerance due to an enhanced ABA induced stomatal closure. XERICO controls the level of ABA by enhancing the transcription of the key ABA biosynthetic gene AtNCED3. The findings indicate that the protein degradation mediated by the ubiquitin/proteasome pathway plays a fundamental role in ABA homeostasis and response.

Sumoylation was also reported to participate in responses to phosphate starvation, and to the tolerance to low and high temperatures. An increase in the levels of SUMO-protein conjugates was also detected in water-deprived plants.

The concerted actions of the transcriptional, post-transcriptional and post-translational mechanisms ensures temporally and spatially appropriate patterns of downstream gene expression and ultimately the shaping of transcriptome and proteome of stress-exposed plants to switch on adaptive response. The complete understanding of the interplay of these three regulatory systems is crucial for the understanding of the molecular mechanisms governing plant adaptation to environment as well as for plant improvement for stress tolerance.

## miRNAs in Plant Responses to Abiotic Storess — An Additional Post-transcriptional Regulation Layer may Apply

Plant responses to abiotic stress such as water deficit involve an intricate regulation of gene expression at the transcriptional and post-transcriptional levels. MicroRNAs (miRNAs) are a class of small non-coding RNAs molecules (21-24 nt) involved in post-transcriptional regulation of gene expression. miRNAs were shown to be involved in plant development, biotic and abiotic stress responses.

In plants, microRNAs repress gene expression by directing mRNA degradation or translational arrest: miRNAs guide Argonaute (AGO) proteins to bind to matching target mRNAs in a RNA-induced silencing complex (RISC), promoting cleavage of mRNAs with near perfect base complementarity and inhibiting translation of those with lower complementarity.

The first reports assigning miRNAs to have a role in shaping plant responses to abiotic stresses were based on small RNA cloning and sequencing, complemented with analyses of miRNA expression profiles and miRNA target prediction. Since then, the application of high-throughput sequencing technology and genomic approaches like microarray analyses to evaluate the profile of miRNA expression in various tissues and conditions, associated to improved bioinformatic tools to identify miRNAs and their targets, have allowed an extensive recognition of stress-responsive small RNAs and their targets in various plant species.

Sequencing of miRNAs in Legumes was first reported in Medicago truncatula and Glycine max but there are references to small RNAs in other Legumes back to 2004, with a size population of small RNA molecules being identified in the phloem sap of Lupinus albus. These findings were the basis of a systemic signalling mechanism in which small RNAs movement is facilitated by chaperone proteins to exert their action at a distance.

One of the most extensively studied miRNAs in the context of abiotic stresses have been the miRNAs involved in nutrient deprivation miR395, miR398 and mir399, all identified in the phloem sap of nutrient deprived plants. In fact, studies in Arabidopsis have established that miR395 (sulphate), miR399 (phosphate) and miR398 (copper) regulate these nutrients homeostasis by moving along the phloem to inform the roots of the nutrient status of the shoot.

The miRNA395 gene-family targets genes involved in sulphate translocation (the low-affinity transporter SULTR2;1) and assimilation (the ATP sulphurylases, APS). Importantly, miR395 itself is regulated by a transcription factor, the SULFUR LIMITATION 1 (SLIM1). The miR395/APS-SULTR2;1/SLIM1 regulatory module is involved in root-to-shoot sulphate translocation as a strategy to improve sulphate assimilation in the leaves during sulphate starvation.

The miR399 gene-family is strongly and specifically induced by inorganic phosphate limitation in the shoot and targets PHO2, an E2 ubiquitin-conjugating enzyme that represses Pi uptake. As for miR395, also the expression of miR399 is regulated by a transcription factor, the MYB TF PHOSPHATE STARVATION RESPONSIVE1 (PHR1;). The miR399/PHO2/PHR1 regulatory module operates under Pi deprivation: miR399 is induced by PHR1 in the leaves, travels along the phloem to repress PHO2 expression in the roots thereby releasing several protein targets from ubiquitinylation-dependent degradation, including transporters involved in Pi allocation inside the plants and increasing Pi content in the shoot. A worth mentioning aspect of the miR399 regulatory module is the extra layer of miR399 activity regulation exerted by IPS1 (induced by phosphate starvation1). IPS1 is a non-protein coding transcript with sequence complementarity to miR399 that sequesters miR399 thus inhibiting its repressing activity over its target. This mechanism designated as target mimicry was first described in plants and more recently discovered in animals and expands the regulatory post-transcriptional gene expression network in which miRNAs are involved.

The miR398 (and miR408) are induced by copper limitation and target genes encondig copper proteins like Copper/Zinc superoxide dismutases, cytochrome c oxidase and plantacyanin. Similar to miR395 and miR399, also miR398 and miR408 are regulated by a transcription factor, the SQUAMOSA promoter binding protein–like7 (SPL7) that regulates the expression of several copper-responsive genes. Copper in contrast to sulphate and phosphate is a micronutrient but still the regulation of this nutrient homeostasis is basically similar, as it involves sistemic signalling, a well established regulatory module involving a transcription factor, the miRNA and its target.

The miR395, miR399 miR398 and miR408 were identified in M. truncatula by sequencing libraries of small RNAs from the aerial part. Homologs of known miRNA target genes were identified, such as low affinity sulphur transporter for miR395, COX5b (subunit 5b of mitochondrial cytochrome c oxidase) for miR398, PHO2 for miR399 or plantacyanin for miR408. However, our computational prediction identified many hypothetical genes for miRNA targeting, rendering experimental confirmation a laborious and unsuccessful task.

Some miR398 and miR408 predicted targets were validated by 5'RACE and miR398 and miR408 expression was further investigated in different plant parts and in specific water deficit conditions, showing up-regulation in water deprivation and concomitant down-regulation of their validated targets. These targets were further confirmed by deep sequencing of cleaved miRNA targets (Parallel Analysis of RNA Ends - PARE) in M. truncatula in collaboration with the Tamas Dalmay laboratory.

Still, the bioinformatic prediction of many hypothetical genes for miRNA targeting raises the question whether we are dealing with true or instead pseudo targets and can have a strong implication on our assumptions about the mechanisms of miRNA functioning as they impose an additional layer of post-transcriptional regulation.

Seitz proposed that many computational identified miRNA targets are indeed pseudotargets that prevent miRNAs from binding their true targets by sequestering them. They would have the basic features of miRNA targets identified by the target prediction algorithms: complementarity to miRNAs and phylogenetic conservation but are instead modulators of miRNA expression.

These pseudotargets occur naturally in plants and animals and were firstly associated to miRNA regulation of nutrient deprivation but their involvement in other abiotic stress conditions like water deprivation may also be envisaged.

A 5-year EU FP7 project designated "ABStress - Improving the resistance of legume crops to combined abiotic and biotic stress" was recently started.

## Transgenic Approaches to Improve Abiotic Stress Resistance

The advance in genetic engineering offers new ways to understand the genetic mechanisms of stress-related genes and their contribution to the plant performance under stress. However, while a great degree of success has been obtained in the production of herbicide, virus and fungal-resistant plants and plants with fortified nutritional values using transgenic tools, the same has not been the case in production of abiotic stress-tolerant crops. This is largely due to the complex genetic mechanisms that govern abiotic stress tolerance. Additionally, as previously referred, in natural conditions, crops can suffer from different stress combinations, at different development stages and during different time periods.

Recently, several reviews were published concerning genetic engineering for abiotic stress tolerance, most focused in model but also in crop plants (e.g.). Possible targets for genetic engineering towards abiotic stress in plants are genes belonging to structural and regulatory categories. They can be modified (for example truncated) and fused to other genetic components such as signal peptides that direct their expression to specific organelles and reporter genes for early detection in transgenic plants. After the proper cloning of the desired genes, they are engineered for their expression to be regulated in a time and space context, using specific promoters. The approach can take into account if it is desirable to have the gene expression upregulated, by sense overexpression of the transgene, or downregulated, by the antisense or RNA interference (RNAi) techniques.

Presently, numerous genes associated to plant responses to abiotic stress have been identified and characterized in laboratory studies. Engineered overexpression of biosynthetic enzymes for osmoprotectants such as glycine betaine; stress induced proteins such as LEA proteins; scavengers of reactive oxygen species; transcription factors or signal transduction components were reported. Since stress resistance is a complex trait regulated by several genes acting in a concerted way during the process, it is not surprising that transgenic approaches using a single stress-related gene will only lead to marginal stress improvement. One of the major challenges is the introduction of multiple genes by pyramiding strategies or co-transformation.

It is also expected that several areas, such as post-transcriptional regulation involving protein modification, protein degradation and RNA metabolism will emerge. An example is the application of miRNAs in the improvement of stress resistance. The discovery of miRNAs involved in the regulation of stress responses and discovering the potential use of these miRNAs to modulate or even increase stress resistance in plants is an open field of research. As an example, Sunkar and co-workers have generated transgenic Arabidopsis thaliana plants overexpressing a miR398-resistant form of a plastidic Cu/Zn Super Oxide Dismutase (Cu/Zn-SOD;CSD2) and confirmed that transgenic plants accumulate more CSD2 mRNA than plants overexpressing a regular CSD2 and are consequently much more tolerant to high light, heavy metals, and other oxidative stresses. These results suggest that understanding posttranscriptional gene regulation is important to widen our ability to manipulate stress tolerance in plants and offer an improved strategy to engineer crop plants with enhanced stress tolerance.

The process of generating transgenic lines requires success in the transformation method and proper incorporation of stress resistance genes into plants. The most used method to transfer foreign genes into plant cells and the subsequent regeneration of transgenic plants is based on the natural system, the Agrobacterium-mediated plant transformation. Particle bombardment has also been exploited extensively for plant transformation especially in species recalcitrant to Agrobacterium infection such as maize. The development of new plant transformation vectors namely using new-plant associated bacteria (such as from the Rhizobiacea family) has also proved to be an effective approach to generate transgenic plants from explants/genotypes unsuitable for Agrobacterium-mediated transformation methodology.

The promoters that have been most commonly employed in the production of abiotic stress-tolerant plants include the cauliflower mosaic virus (CaMV) 35S promoter (mostly used for dicot crops) and the actin 1 promoter (used for expression of transgenes in monocot crops). As these promoters are constitutive, the downstream transgenes are expressed in all organs and at all stages which is unnecessary as well as demanding on the energy reserves of the cell. In some cases, constitutive expression of a gene normally only induced by stress can have negative effects on growth and development when stress is not present (pleiotropic effects). The use of inducible promoters that allow the expression of a transgene only when it is required could therefore be the ideal solution. There is a strong need to obtain an increased array of inducible promoters, which are expressed only when exposed to stress situations, and to pair such promoters with the stress tolerance-related genes in the adequate cloning vectors. Additional tests need to be performed to guarantee that obtained stress-inducible promoters work in heterologous plant systems.

Concerning the improvement of stress resistance, the past decade has witnessed the utilization of transgenic approaches for experimental purposes, mainly in model plant systems but not in important agricultural species or crops. Nevertheless, the creation of stress-tolerant crops either by genetic engineering or through conventional breeding has covered almost all aspects of plant science, and is pursued by both public and private sector researchers. One of the major goals of transgenic technology is to produce plants not only able to survive stress, but also capable to grow under adverse conditions with substantial biomass production, thus overcoming the negative correlation between drought resistance traits and productivity, which was often present in past breeding programs. In the case of crop plants, it is ultimately the yield of genetically altered plants under specific field conditions that will determine whether or not a specific gene, or metabolic or signaling pathway, is of technologic importance. One successful case in releasing tolerant plants to abiotic stresses is the transgenic maize line resistant to drought developed by the Monsanto company. This maize line (MON87460) was recently approved in the USA and is able to growth in soils with reduced water content due to the presence of a cold shock protein – CSPB from Bacillus subtilus.

## Engineering Trehalose Accumulation

Trehalose is a disaccharide, containing two glucose molecules. Trehalose was first discovered in 1832 from the Ergot of rye and since then isolated from numerous organisms, including algae, fungi, bacteria, insects and crustaceans. Trehalose is nevertheless considered non-occurring in measurable amounts in plants, with the exception of a few species, notably the so called "resurrection plants", able of surviving the loss of most of their water content until a quiescent stage is achieved and upon watering rapidly revive and restored to their former state.

Trehalose can be synthesized by three different pathways and the most frequent in nature involves the enzyme trehalose-6-phosphate synthase (TPS; EC 2.4.1.15) that catalyzes the transfer of glucose from UDP-glucose to glucose-6-phosphate to produce trehalose-6-phosphate plus UDP. Another enzyme, trehalose-6-phosphate phosphatase (TPP; EC 3.1.3.12) converts trehalose-6-phosphate to free trehalose. Genes codifying both enzymes have been isolated in several species including Sacharomyces cerevisiae and Escherichia coli and several plant species such as Arabidopsis and rice. Trehalose may be degraded by the enzyme trehalase (EC 3.2.1.28).

In living organisms, several functional properties have been proposed for trehalose: energy and carbon reserve, protection from dehydration, protection against heat, protection from damage by oxygen radicals and protection from cold. As trehalose, sucrose is one of the few free disaccharides in nature. Both are non-reducing sugars and synthesized by similar pathways. Contrary to trehalose, sucrose synthesis is mainly limited to photosynthetic organisms, where it holds a central position as the major product of photosynthesis and as a transport molecule involved in growth, development, storage, signal transduction and acclimation to environmental stress. Sucrose transport is finally energetically superior to trehalose transport making it more "preferred" to plants metabolism. It is hence often suggested that trehalose is evolutionary more ancient than sucrose.

As trehalose is present in so low or in undetectable amounts in most plants, it is unlikely that under natural conditions and with the exception of desiccation tolerant plants, this sugar might play a role in stress protection in plants. Nevertheless, other roles have been proposed for trehalose and trehalose-6-phosphate synthase: regulation of plant growth and development; broad spectrum agent preventing symbiosis between susceptible plants and trehalose producing microorganisms; the regulation of carbohydrate metabolism or the perception of carbohydrate availability; the regulation of embryo maturation; implication on vegetative growth and transition to flowering; implication on seedling development; and regulation of glucose, abscisic acid and stress signaling. Trehalose plays several roles in carbohydrate metabolism, with a number of processes and pathways being affected.

For all that was stated above, trehalose is one of the most studied osmoprotectants and in recent years there has been a growing interest in trehalose metabolism as a means of engineering stress tolerance in crop plants. Several experiments have been conducted to obtain transgenic plants over-expressing genes codifying enzymes of the trehalose biosynthetic pathway of E. coli and S. cerevisiae, using both model plants like tobacco (Nicotiana tabacum) and crop plants such as potato (Solanum tuberosum), rice (Oryza sativa) and more recently tomato (Lycopersum esculentum). Additional, attempts have been made using an alternative approach: the inhibition of the expression of trehalase gene.

Generally speaking, transgenic plants were found to have higher tolerance than controls to some form of water stress imposed, following in most cases, confirmed trehalose accumulation. Albeit such fact, trehalose engineered plants frequently had altered phenotypes, particularly dwarfism and leaf abnormalities. Such fact was particularly true for the first transformation events in which genes of microbial origin were used. Later events, in which endogenous or plant origin genes were used seem to counter that tendency. Genetic engineering of plants with trehalose biosynthesis genes seems therefore to be of extreme pertinence to the increase of abiotic stress tolerance in plants, particularly plants of agricultural importance such as cereals and legumes.

## Engineering Polyamine Accumulation

Polyamines (PAs) are small (low-molecular-weight), positively charged, aliphatic amines that are found in all living organisms. The major forms of PAs are putrescine (Put), spermidine (Spd) and spermine (Spm), although plants also synthesized a variety of other related compounds. Arginine (Arg) and ornithine (Orn) are the precursors of plant PAs. Ornithine decarboxylase (ODC; EC 4.1.1.17) converts Orn directly into Put. The other biosynthetic route to Put, via arginine decarboxylase (ADC; EC 4.1.1.19), involves the production of the intermediate agmatine (Agm) followed by two successive steps catalysed by agmatine iminohydrolase (AIH, EC 3.5.3.12) and N-carbamoylputrescine amidohydrolase (CPA, EC 3.5.1.53). In animals and fungi Put is synthesized primarily through the activity of ODC while in plants and bacteria the main pathway involves ADC. Aminopropyl groups, donated by decarboxylated S-adenosyl methionine (dcSAM), must be added to convert Put into Spd and Spm in a reaction catalysed by spermidine synthase (SPDS; EC 2.5.1.16) and spermine synthase (SPMS; EC 2.5.1.22), respectively. Polyamines levels in plants increase under a number of environmental stress conditions, including drought and salinity. Several biological roles were proposed for polyamines action in stress situations; PAs could act as osmoprotectants, as scavengers of active oxygen species (AOS) or by stabilizing cellular structures, such as thylakoid membranes. The first reports of transgenic approaches using genes responsible for PA biosynthesis were conducted in two species, tobacco and rice. Recently, new insights into the role and regulatory function of polyamines in plant abiotic stress tolerance have been achieved, with several abiotic (salt, drought, freezing, heat) stress tolerant transgenic plants overproducing polyamines.

Among abiotic stresses drought is the main abiotic factor as it affects 26% of arable area. Plants respond to changes in water status by accumulating low molecular-weight osmolytes including PAs. Polyamines may have a primary role of turgor maintenance but they may also be involved in stabilizing proteins and cell structures. The polycationic nature of PAs at physiological pH is believed to mediate their biological activity, since they are able to bind to several negatively charged molecules, such as DNA, membrane phospholipids, pectic polysaccharides and proteins.

In respect to the antioxidant activity of PAs, the research data is contradictory; on the one hand, PAs have been suggested to protect cells against AOS and on the other hand, their catabolism generates AOS. PA catabolism produces $H_2O_2$, a signaling molecule that can act promoting activation of antioxidative defense response upon stress, but can also act as a peroxidation agent. The effect of increased putrescine (Put) accumulation was found to negatively impact the oxidative state of poplar cells in culture due to the enhanced turnover of Put. Gill and Tuteja stated that, while increase Put accumulation may have a protective role against AOS in plants, enhanced Put turnover can actually make them more vulnerable to increased oxidative damage. The higher polyamines, Spd and Spm are believed to be most efficient antioxidants and are considered scavengers of oxyradical.

As plants with elevated putrescine contents are able to tolerate drought stress because Put has a direct protective role in preventing the symptoms of dehydration, higher PAs (Spd and Spm) appear to play an important in role in stress recovery. Recently, transgenic rice plants overexpressing samdc (S-Adenosyl methionine decaboxylase gene), with increased Spd and Spm levels, were considered to be non drought tolerant, but showed a more robust recovery from drought compared to wild type. The de novo synthesis of Spd and Spm in transgenic plants under drought stress, at the expenses of Put, was responsible for the stress tolerance observed in these plants.

The covalent linkage of PAs to proteins appeared to be of extreme importance in plant light-induced stabilization of the photosynthetic complexes and Rubisco therefore exerting a positive effect on photosynthesis and photo-protection. Also in the cytosol, they are involved, mediated by transglutaminase (TGase) activity, in the modification of cytoskeletal proteins and in the cell wall construction/organization. The characterization at the proteomic level of the TGase interaction with thylakoid proteins, demonstrated its association with photosystem II (PSII) protein complexes using maize thylakoid protein extracts. Binding of Put to thylakoid membranes has been proposed to be a photoadaptation response under controlled stress conditions. Campos and collaborators results reinforce the importance of the TGase in photo-protection by polyamine conjugation to light-harvesting complex II (LHCII) proteins.

Recently, PAs were proposed to be components of signaling pathways and fulfill the role of second messengers. Studies with ABA-deficient and ABA-insensitive Arabidopsis mutants with differential abiotic stress adaptations support the conclusion that the up-regulation of PA biosynthetic genes and Put accumulation under water stress are mainly ABA-dependent responses. To reinforce the fact that PAs biosynthesis may be regulated by ABA, several stress-responsive elements, like drought responsive (DRE), low temperature-responsive (LTR) and ABA-responsive elements (ABRE and ABRE-related motifs) are present in the promoters of the polyamine biosynthetic genes. Liu et al. also found that inward potassium channels were targets for PA regulation of stomatal movements. Since ABA signaling pathway in stomata regulation involves many different components including signaling molecules like AOS, $IP_3$, $Ca^{2+}$ and nitric oxide (NO), evidences point to an interplay between ABA, polyamines, $H_2O_2$ and NO in stomata regulation.

We transformed the model legume Medicago truncatula cv. Jemalong with the arginine decarboxylase gene (adc) from Avena sativa to overexpress the heterologous ADC enzyme aiming to increase the levels of polyamines in transgenic plants. Several transgenic lines overexpressing This oat adc construct were obtained. The oat adc cDNA under the control of a CaMV 35S constitutive promoter was previously transferred into rice plants and those authors found increased Put levels in regenerated plants and observed minimized chlorophyll loss during drought stress. However, constitutive over-expression of this gene severely affected developmental patterns of those plants. Afterwards, the same group used the monocot maize's ubiquitin-1 (Ubi-1) promoter to overexpress the Datura adc gene and found that transgenic plants, with increased Put levels, were tolerant to drought stress. The Ubi-1 promoter is known to contain a number of stress-responsive elements that enhance transgene expression under drought stress and hence function as a stress-inducible promoter. Roy and Wu also found that the expression of the adc transgene under the control of an ABA-inducible promoter led to stress-induced upregulation of ADC activity and polyamine accumulation in transgenic rice plants. Second-generation transgenic rice plants showed an increase in biomass under salinity–stress conditions.

In our M. truncatula system, no altered external morphology was observed in adc transgenic plants, that were successfully developed without phenotypic visible alterations and produced seeds ($T_2$ generation). One specific transgenic line (L108) expressing the heterologous adc transgene had a very high accumulation of Agmatine (22-fold) (the direct product of the ADC enzyme and intermediate in the Put biosynthesis) and moderately related increase of Put (1.7-fold) and Spd (1.9-fold) levels, compared to control plants. These results are consistent with several reports that suggest PAs levels are under strict homeostatic regulation.

Nevertheless, several recent studies have concluded on the feasibility of PA biosynthesis engineered for the production of stress-tolerant plants. Accumulating experiments and their main results are summarized in Table. The constitutive expression of homologous adc1 and adc2 in Arabidopsis resulted in freezing and drought tolerance, respectively; with a patent application for "Plant resistance to low-temperature stress and method of production thereof" by. In another work, transgenic tomato lines transformed with the yeast samdc fused with a ripening-specific promoter E8, over-accumulate Spd and Spm and, interestingly, showed phenotypes of agronomical importance such as enhanced phytonutrient content and fruit quality. Polyamine-accumulating transgenic eggplants exhibited increased tolerance to multiple abiotic stresses (salinity, drought, low and high temperature and heavy-metal) and also biotic resistance against fungal disease caused by Fusarium oxysporium. These authors used a construct similar to ours, with the adc gene from oat under the control of the constitutive CaMV 35S promoter and found that some transgenic eggplants lines showed an enhanced level of Put, Spd and in some cases also Spd. These lines also showed increase in ADC and also on the activity of the PA catabolic enzyme, diamine oxidase (DAO).

There are several reports in which the plant response to diverse abiotic stress is associated to the stimulation of polyamine oxidation. However, the precise role of polyamine catabolism in the plant response to environmental stress remains elusive. Considering these results, further research concerning the PAs changes and the global response of our M. truncatula diverse germplasm with altered PA content to multiple stresses should be developed in the near future.

Table: Genetic Engineering of plants towards polyamine accumulation.

| Gene/Promoter | Origin | Plant | Main Effects |
|---|---|---|---|
| odc; CaMV 35S | S. cerevisae | Tobacco | Increased ODC activity; Increased Put and Nicotine |
| samdc; CaMV 35S | Human | Tobacco | Increased SAMDC activity; Spd and Spm levels. Lower Put levels. Thick leaves, stems and stunting. |
| adc; Tet-inducible promoter | Oat | Tobacco | Increased ADC activity; Phenotypic alterations pp to Put levels (thin stems and leaves, leaf necrosis, chlorosis, short internodes and growth inhibition) |
| adc; CaMV 35S | Oat | Tobacco | Increased ADC activity; ODC and SAMDC normal; Increased Agm; Put, Spd and Spm normal. |
| adc; CaMV 35S | Oat | Rice | Increased Put and less chlorophyll loss during drought. Severe altered phenotypes. |
| adc; ABA-inducible promoter | Oat | Rice | Increased Put, ADC activity and biomass under salt stress. |
| samdc; E8 promoter | Yeast | Tomato | Increased Spd and Spm. Enhanced phytonutrient content and fruit quality |
| adc; Ubi-1 promoter | D. stramonium | Rice | Higher Put, Spd and Spm levels and drought tolerance |
| adc; CaMV 35S | Oat | M. truncatula | Increased Agm, Put and Spd levels. Absence of phenotypic alterations and altered growth (second generation homozygous plants). |
| adc; CaMV 35S | Oat | Eggplant | Increased Put, Spd and Spm levels; multiple abiotic stress resistance and fungal resistance. |
| adc1; adc2; CaMV 35S | Arabidopsis | Arabidopsis | Increased Put; freezing and drought tolerance. |

## Engineering Accumulation of Photo-protective Proteins — ELIPs

To cope with environmental stresses, plants activate a large set of genes, which lead to the accumulation of specific stress-associated proteins. The stomatal limitation on photosynthesis imposed by the earlier stages of water deficit (WD) result in a decrease of primary electron acceptors available for photochemistry. If protection mechanisms are not activated, the excess of absorbed energy may induce photo-oxidative damage in chloroplast structures.

ELIPs and ELIP-like proteins are pigment-binding components of the thylakoid membrane widely distributed among plant species and belong to the chlorophyll a/b-binding protein (cab) family). ELIPs are widely present among different plant species like pea, barley,

Craterostigma plantagineum, Dunaliella bardawil, Sporobolus stapfianus, Arabidopsis thaliana, Tortura ruralis, Nicotiana tabacum and recently found in Coffea canephora.

Contrary to the other members of the cab family that are expressed constitutively, ELIPs accumulate transiently during the greening of etiolated plants and in developing plastid membranes. In mature plants, ELIPs also accumulate in response to various stress conditions including ABA or desiccation, nutrient starvation, high light, UV-B, cold, methyl jasmonate, salinity and senescence. ELIPs and ELIP-like proteins are thought to protect the chloroplast apparatus from photooxidation by: (a) acting as transient pigment-binding proteins during biogenesis or turnover of chlorophyll binding proteins; (b) binding or stabilising carotenoids like zeaxantin and lutein; (c) stabilising the pigment-protein complexes and favouring their appropriate assembly; (d) dissipating the excessive absorved light energy at the reaction center of the PSII, in the form of heat or fluorescence.

# EFFECTS OF ABIOTIC STRESS

Recent advances in biotechnology have dramatically changed our capabilities for gene discovery and functional genomics. For the first time, we can now obtain a holistic “snapshot” of a cell with transcript, protein and metabolite profiling. Such a “systems biology” approach allows for a deeper understanding of physiologically complex processes and cellular function. New models can be formed from the plethora of data collected and lead to new hypotheses generated from those models.

Understanding the function of genes is a major challenge of the post-genomic era. While many of the functions of individual parts are unknown, their function can sometimes be inferred through association with other known parts, providing a better understanding of the biological system as a whole. High throughput omics technologies are facilitating the identification of new genes and gene function. In addition, network reconstructions at the genome-scale are key to quantifying and characterizing the genotype to phenotype relationships.

## Plant Responses to the Environment are Complex

Plants are complex organisms. It is difficult to find an estimate of the total number of cells in a plant. Estimates of the number of cells in the adaxial epidermal layer and palisade mesophyll of a simple Arabidopsis leaf are approximately 27,000 and 57,000 cells, respectively. Another estimate of the adaxial side of the epidermal layer of the 7$^{th}$ leaf of Arabidopsis was close to 100,000 cells per cm$^2$ of leaf area. An Arabidopsis plant can grow as large as 14 g fresh weight with a leaf area of 258 cm$^2$ (11 g fresh weight). Thus, we estimate that a single Arabidopsis plant could have approximately 100 million cells (range of 30 to 150 million cells assuming 2.4 to 11 million cells per g fresh weight.)

A one million kg. redwood tree could possibly have 70 trillion cells assuming a cell size 100 times larger than an Arabidopsis cell. Combine that with developmental changes, cell differentiation and interactions with the environment and it is easy to see that there are an infinite number of permutations to this complexity.

There is additional complexity within the cell with multiple organelles, interactions between nuclear, plastidial and mitochondrial genomes, and between cellular territories that behave like symplastically isolated domains that are able to exchange transcription factors controlling gene expression and developmental stages across the plasmodesmata. A typical plant cell has more than 30,000 genes and an unknown number of proteins, which can have more than 200 known post-translational modifications (PTMs). The molecular responses of cells (and plants) to their environment are extremely complex.

## Environmental Limits to Crop Production

In 1982, Boyer indicated that environmental factors may limit crop production by as much as 70%. A 2007 FAO report stated that only 3.5% of the global land area is not affected by some environmental constraint. While it is difficult to get accurate estimates of the effects of abiotic stress on crop production, it is evident that abiotic stress continues to have a significant impact on plants based upon the percentage of land area affected and the number of scientific publications directed at various abiotic stresses table. If anything the environmental impacts are even more significant today; yields of the "big 5" food crops are expected to decline in many areas in the future due to the continued reduction of arable land, reduction of water resources and increased global warming trends and climate change.

| Stress Type | % of global land area affected* | % of global rural land area affected** | Number of Publications*** |
|---|---|---|---|
| Abiotic Stress | | 96.5 | 35,363 |
| Water | | | 4819 |
| Deficit or Drought | 64 | 16 | 4137 |
| Flooding or Anoxia | 13 | 10 | 682 |
| Temperature | | | 9715 |
| Cold | 57 | 26 | 3798 |
| Chilling | | | 187 |
| Freezing | | | 350 |
| High or heat | | | 5380 |
| Light | | | 7659 |
| Low | | | 3081 |
| High | | | 4578 |
| Chemical/Soil | | 50 | 12391 |
| Salt or salinity | 6 | 6 | 3498 |

| | | | |
|---|---|---|---|
| Mineral deficiency or low fertility | 9 | 39 | 222 |
| Mineral toxicity | | | 437 |
| Acid soil | 15 | | 3646 |
| Air pollutants | | | |
| Ozone | | | 1369 |
| Sulfur dioxide | | | 378 |
| $NO_x$ oxide | | | 2001 |
| Elevated $CO_2$ | | | 840 |
| Miscellaneous (e.g. wind, mechanical, etc.) | | | 779 |

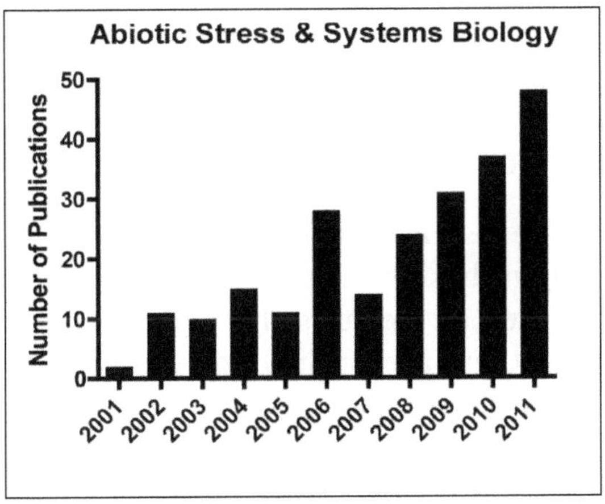


The number of publications per year related to systems biology and abiotic stress. Key words used in the search of PubMed included: plant, systems biology, and abiotic stress (including stress sub-terms; e.g. drought or water deficit or dehydration).

## Multiple Factors Limit Plant Growth

Fundamentally, plants require energy (light), water, carbon and mineral nutrients for growth. Abiotic stress is defined as environmental conditions that reduce growth and yield below optimum levels. Plant responses to abiotic stresses are dynamic and complex, they are both elastic (reversible) and plastic (irreversible).

The plant responses to stress are dependent on the tissue or organ affected by the stress. For example, transcriptional responses to stress are tissue or cell specific in roots and are quite different depending on the stress involved. In addition, the level and duration of stress (acute vs chronic) can have a significant effect on the complexity of the response.

Water deficit inhibits plant growth by reducing water uptake into the expanding cells, and alters enzymatically the rheological properties of the cell wall; for example, by the activity of ROS (reactive oxygen species) on cell wall enzymes. In addition, water deficit alters the cell wall nonenzymatically; for example, by the interaction of pectate and calcium. Furthermore, water conductance to the expanding cells is affected by aquaporin activity and xylem embolism. The initial growth inhibition by water deficit occurs prior to any inhibition of photosynthesis or respiration.

The growth limitation is in part due to the fundamental nature of newly divided cells encasing the xylem in the growing zone. These cells act as a resistance to water flow to the expanding cells in the epidermis making it necessary for the plant to develop a larger water potential gradient. Growth is limited by the plant's ability to osmotically adjust or conduct water. The epidermal cells can increase the water potential gradient by osmotic adjustment, which may be largely supplied by solutes from the phloem. Such solutes are supplied by photosynthesis that is also supplying energy for growth and other metabolic functions in the plant. With long-term stress, photosynthesis declines due to stomatal limitations for $CO_2$ uptake and increased photoinhibition from difficulties in dissipating excess light energy.

One of the earliest metabolic responses to abiotic stresses and the inhibition of growth is the inhibition of protein synthesis and an increase in protein folding and processing. Energy metabolism is affected as the stress becomes more severe (e.g. sugars, lipids and photosynthesis). Thus, there are gradual and complex changes in metabolism in response to stress.

## Central Regulators Limit Key Plant Processes

The plant molecular responses to abiotic stresses involve interactions and crosstalk with many molecular pathways. Systems biology and omics approaches have been used to elucidate some of the key regulatory pathways in plant responses to abiotic stress.

One of the earliest signals in many abiotic stresses involve ROS and reactive nitrogen species (RNS), which modify enzyme activity and gene regulation. ROS signaling in response to abiotic stresses and its interactions with hormones has been thoroughly reviewed. ROS and RNS form a coordinated network that regulates many plant responses to the environment; there are a large number of studies on the oxidative effects of ROS on plant responses to abiotic stress, but only a few studies documenting the nitrosative effects of RNS.

Hormones are also important regulators of plant responses to abiotic stress. The two most important are abscisic acid (ABA) and ethylene. ABA is a central regulator of many plant responses to environmental stresses, particularly osmotic stresses. Its signaling can be very fast without involving transcriptional activity; a good example is the control of stomatal aperture by ABA through the biochemical regulation of ion and water transport processes. There are slower responses to ABA involving transcriptional responses that regulate growth, germination and protective mechanisms.

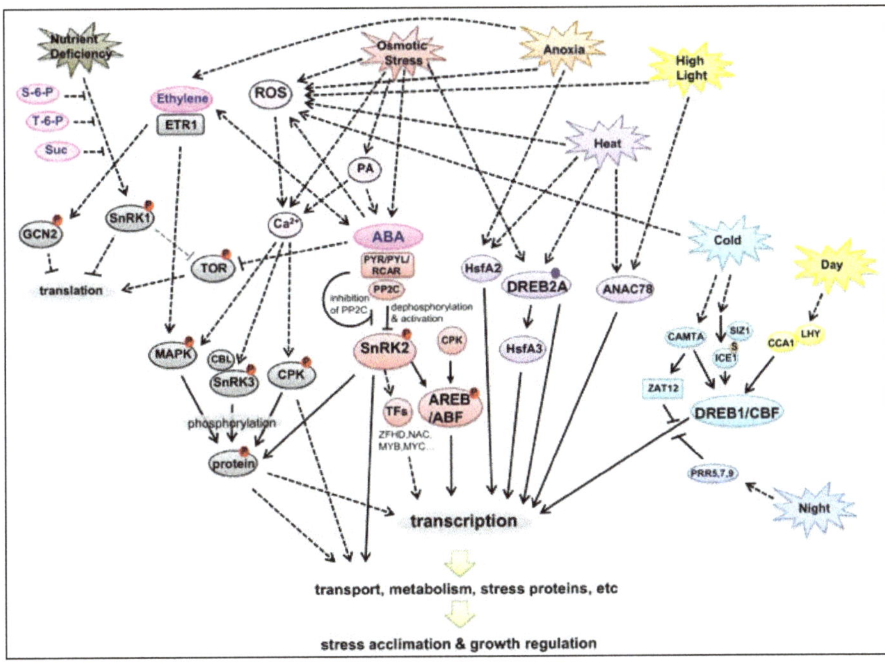


A simplified working model of a signaling network of plant responses to abiotic stress. Ovals represent proteins, metabolites or processes. Metabolites have magenta color. Phosphorylated proteins have red circles with a P inside. Sumoylated protein has an orange circle with an S inside. The solid purple circle indicates that DREB2 needs modification to be activated. Solid lines represent direct connections; dotted lines represent indirect connections (acting through some intermediate molecule). The gray line indicates that this reaction has not been shown in plants. Not all linkages and details of stress and hormone effects are shown in this diagram in order to simplify the model. Abbreviations: ABA (abscisic acid), ANAC (Arabidopsis NAC domain-containing protein), CAMTA (calmodulin-binding transcription activator), CBL (calcineurin B-like interacting protein kinase), CCA (circadian clock associated), CPK (calcium-dependent protein kinase), DREB/CBF (dehydration response element binding protein/C-repeat binding factor), ETR1 (ethylene response 1), GCN2 (general control non-repressible 2), HSF (heat shock factor), ICE (inducer of CBF expression), MAPK (mitogen-activated protein kinase), LHY (late elongated hypocotyl), PA (phosphatidic acid), PP2C (protein phosphatase 2C), PRR (pseudo response regulator), PYR/PYL/RCAR (ABA receptors), RNS (reactive nitrogen species), ROS (reactive oxygen species), SIZ (SAP and Miz domain protein), SnRK (sucrose nonfermenting-1 related kinase), TFs (transcription factors), TOR (target of rapamycin), ZAT (zinc finger protein).

Recently, the essential components of ABA signaling have been identified, and their mode of action was clarified. The current model of ABA signaling includes three core components, receptors (PYR/PYL/RCAR), protein phosphatases (PP2C) and protein kinases (SnRK2/OST1). The PYR/PYL/RCAR proteins were identified as soluble ABA receptors by two independent groups. The 2C-type protein phosphatases (PP2C) including ABI1 and ABI2, were first identified from the ABA-insensitive Arabidopsis mutants abi1-1 and abi2-1, and they act as global negative regulators of ABA signaling. SNF1-related protein kinase 2 (SnRK2) is a family of protein kinases isolated as ABA-activated protein kinases. In Arabidopsis, three members of this family, SRK2D/SnRK2.2, SRK2E/OST1/SnRK2.6, and SRK2I/SnRK2.3, regulate ABA signaling positively and globally, as shown in the triple knockout mutant srk2d srk2e srk2i (srk2dei)/snrk2.2 snrk2.3 snrk2.6, which lacks ABA responses. The PYR/PYL/RCAR - PP2C - SnRK2 complex plays a key role in ABA perception and signaling.

Studies of the transcriptional regulation of dehydration and salinity stresses have revealed both ABA-dependent and ABA-independent pathways. Cellular dehydration under water limited conditions induces an increase in endogenous ABA levels that trigger downstream target genes encoding signaling factors, transcription factors, metabolic enzymes, and others. In the vegetative stage, expression of ABA-responsive genes is mainly regulated by bZIP transcription factors (TFs) known as AREB/ABFs, which act in an ABA-responsive-element (ABRE) dependent manner. Activation of ABA signaling cascades result in enhanced plant tolerance to dehydration stress. In contrast, a dehydration-responsive cis-acting element, DRE/CRT sequence and its DNA binding ERF/AP2-type TFs, DREB1/CBF and DREB2A, are related to the ABA-independent dehydration and temperature responsive pathways. DREB1/CBFs function in cold-responsive gene expression, whereas DREB2s are involved in dehydration-responsive and heat-responsive gene expression.

Ethylene is also involved in many stress responses, including drought, ozone, flooding (hypoxia and anoxia), heat, chilling, wounding and UV-B light. Ethylene signaling is well defined. There are known interactions between ethylene and ABA during drought, fruit ripening, and bud dormancy. All of these interactions make the plant response to stress very complex.

In yeast, the well-documented central regulators of protein synthesis and energy are SnRK1 (Snf1/AMPK), TOR1 and GCN2. These proteins are largely controlled by the phosphorylation of enzymes; all three are protein kinases acting as key hubs in the coordination of metabolism during stressful conditions. In plants, TOR activity is inhibited by osmotic stress and ABA and GCN2 activity is stimulated by UV-light, amino acid starvation, ethylene, and cold stress. SnRK1 responds to energy depletion, such as low light, nutrient deprivation or hypoxic conditions, and interacts with both glucose and ABA signaling pathways. One of the results of this coordinated response is the inhibition of protein synthesis.

Many abiotic stresses directly or indirectly affect the synthesis, concentration, metabolism, transport and storage of sugars. Soluble sugars act as potential signals interacting with light, nitrogen and abiotic stress to regulate plant growth and development; at least 10% of Arabidopsis genes are sugar-responsive. Mutant analysis has revealed that sugar signaling interacts with ethylene, ABA, cytokinins, and light. In grapevine, sugar and ABA signaling pathways interact to control sugar transport. An ASR (ABA, stress-, and ripening-induced) protein isolated from grape berries is upregulated synergistically by ABA and sugars, and upregulates the expression of a hexose transporter. VVSK1, a GSK3 type protein kinase, is also induced by sugars and ABA, and upregulates the expression of several hexose transporters.

Stresses such as sugar starvation and lack of light stimulate SnRK1 activity. Suc-P synthase (SPS), 3-hydroxy-3-methylglutaryl-CoA reductase, nitrate reductase, and trehalose-6-P synthase are negatively regulated by SnRK1 phosphorylation, indicating that SnRK1 modulates metabolism by phosphorylating key metabolic enzymes. Post-translational redox modulation of ADPG-pyrophosphorylase, a key control of starch synthesis, by SnRK1 provides an interesting example of interactions between phosphorylation, redox control and sugar metabolism. In Arabidopsis, SnRK1 kinase activity is itself increased by GRIK1 and GRIK2, which phosphorylate a threonine residue of the SnRK1 catalytic subunit. SnRK2 interacts with ABA for the control of stomatal aperture and participates in the regulation of plant primary metabolism. Constitutive expression of SnRK2.6 drastically boosts sucrose and total soluble sugar levels in leaves, presumably by controlling SPS expression.

## Systems Biology Approach to Abiotic Stress

In the post-genomic era, comprehensive analyses using three systematic approaches or omics have increased our understanding of the complex molecular regulatory networks associated with stress adaptation and tolerance. The first one is 'transcriptomics' for the analysis of coding and noncoding RNAs, and their expression profiles. The second one is 'metabolomics' that is a powerful tool to analyze a large number of metabolites. The third one is 'proteomics' in which protein and protein modification profiles offer an unprecedented understanding of regulatory networks. Protein complexes involved in signaling have been analyzed by a proteomics approach. Integration of the different omics analyses facilitates abiotic stress signaling studies allowing for more robust identifications of molecular targets for future biotechnological applications in crops and trees.

## Co-expression Analyses Identify Regulatory Hubs

An important application of transcriptomics data is co-expression analysis of target genes using on-line analytical tools, such as ATTED-II. This approach is very promising for understanding gene-gene correlations and finding master genes in target conditions.

In a series of pioneering papers, Hirai et al. identified MYB transcription factors regulating glucosinolate biosynthesis in Arabidopsis in response to S and N deficiency using an integrated transcriptomics and metabolomics approach. Genes and metabolites in glucosinolate metabolism were found to be coordinately regulated. Co-expression analysis was used to identify two MYB transcription factors that positively regulate glucosinolate metabolism. Then a knock out mutant and ectopic expression of one of the transcription factors was used to validate its positive role in glucosinolate metabolism. Previously unidentified genes were assigned to this biosynthetic pathway and a regulatory network model was constructed.

Mao et al. performed a gene co-expression network analysis of 1094 microarrays of Arabidopsis using a non-targeted approach. They identified 382 modules in this network. The top three modules with the most nodes were: photosynthesis, response to oxidative stress and protein synthesis. Many of the modules also involved responses to environmental stresses. They constructed a cold-induced gene network from a subset of microarrays. The response to auxin stimulus was the most over-represented of the 18 significant modules.

Carrera et al. used the InferGene application to construct a regulatory model of the Arabidopsis genome. They used datasets from 1,486 microarray experiments. Ten genes were predicted to be the most central regulatory hubs influencing the largest number of genes. Included in this set were transcription factor genes involved in auxin (KAN3), gibberellin (MYB29), abscisic acid (MYB121), ethylene (ERF1), and stress responses (ANAC036). They computed the top 12 gene subnetworks; four of these were related to biotic and abiotic stresses. Eighty-five percent of the predicted interactions of the 25% most connected transcription factors were validated in AtRegNet, the Arabidopsis thaliana Regulator Network

Lorenz et al. investigated the drought response of loblolly pine roots and identified a number of hubs in the transcriptional network. Highly ranked hubs included thioredoxin, an inositol transporter, cardiolipin synthase/phosphatidyl transferase, 9-cis-expoxycarotenoid dioxygenase, zeatin O-glucosyltransferase and a SnRK2 kinase. These genes are involved in phospholipid metabolism, ABA biosynthesis and signaling, and cytokinin metabolism; they appear to be important in stress mediation.

Weston et al used weighted co-expression analysis to define six modules for Arabidopsis responses to abiotic stress. Two hubs in the common response module were an ankyrin-repeat protein and genes involved in Ca signaling. They created a compendium of genomic signatures and linked them to their co-expression analysis. Using the same approach, they extended their analyses to the responses of three different plant species to heat and light. Species-specific responses were found involving heat tolerance, heat-shock proteins, ROS, oligosaccharide metabolism and photosynthesis.

## Time-series Analyses Reveal Multiple Phases in Stress Responses

Time-series analyses allow one to distinguish between primary and secondary responses to stress. In a comprehensive time-series transcriptomics analysis of 7 abiotic stresses on different Arabidopsis organs, a core set of genes (50% were transcription factors) of non-specific responses for all stresses were elucidated. Included in this set were the AZF2, ZAT10 and ZAT12 transcription factors. This initial response is thought to be involved in the readjustment of energy homeostasis in response to the stress. With time (after 1 h) more stress-specific profiles developed.

Sun et al. applied a complexity metric to a set of time series data of Arabidopsis with 9 different abiotic stresses. They found that genes with a higher complexity metric had longer 5' intergenic regions and a greater density of cis-regulatory motifs than the genes with a low complexity metric. Many of the cis-regulatory motifs identified were associated with previously characterized stress responses.

Vanderauwera et al. investigated the effects of hydrogen peroxide ($H_2O_2$) signaling during high light stress using microarray analyses. They found that $H_2O_2$ was not only heavily involved in signaling in high light stress, but also salinity, water deficit, heat and cold stress. $H_2O_2$ was a key regulator of small and 70 kD heat shock proteins and many genes of the anthocyanin metabolic pathway. Anthocyanins appear to play an important role as antioxidants in plants. A specific UDP-glycosyltransferase (UGT74E2) was highly regulated by $H_2O_2$. In a subsequent study, UGT74E2 responded quickly to $H_2O_2$ and glycosylated indole-3-butyric acid (IBA) modifying auxin homeostasis, plant morphology and improving stress tolerance to salinity and water deficit. Furthermore, auxin was found to interact with ABA, increasing the ABA sensitivity of the plant. Silencing a poly (ADP-ribose) polymerase improved high light stress tolerance in Arabidopsis. Part of the improved abiotic stress tolerance was ascribed to improved energy-use efficiency and reduced oxidative stress.

Kusano et al. conducted a time-series experiment on the effects of UV-B light on Arabidopsis using both metabolomics and transcriptomics analyses. They found that plants responded in two phases with an upregulation of primary metabolites in the first phase and the induction of protective secondary metabolites, especially phenolics, in the second phase. The induction of phenolics corresponded to transcripts involved in the phenylpropanoid pathway, but the transcripts for primary metabolism were less consistent indicating that this pathway may be regulated by other mechanisms (e.g. kinases).

The transcriptomic response to drought can vary with the time of day. These responses seem to interact with hormonal and other stress pathways that naturally vary during the course of the day. A smaller set of core genes were identified that responded at all times of the day. This set was compared to two previous studies and was whittled down to just 19 genes, including a NF-YB transcription factor, several PP2Cs, a CIPK7, and a sulfate transporter.

Drought stress studies and microarray analyses of three different genotypes of poplar clones grown in two different locations revealed epigenetic regulation to the environment. The tree clones that had a longer history in the environment showed greater changes in DNA methylation, thereby influencing their response to drought.

Shoot tip growth of grapevines was found to be much more sensitive to osmotic stress than gene expression in a time-series experiment of the effects of gradual osmotic stress on grapevine. Proteomics data indicated that changes in protein expression preceded and were not well correlated with gene expression. The integration of transcriptomics data and metabolomics data indicated distinct differences of the responses of salinity and an isosmotic water deficit. Drought-stressed plants induced greater responses in processes needed for osmotic adjustment and protection against ROS and photoinhibition. Salinity induced greater responses in processes involved in energy metabolism, ion transport, protein synthesis and protein fate. A comparison to similar short-term stresses indicated that a gradual, chronic stress response was more complex than an acute stress response.

The effect of water-deficit on Cabernet Sauvignon berries (a red wine grape) in the field was studied using transcriptomics, proteomics and metabolomics. Integrated analyses confirmed that the phenylpropanoid pathway (including anthocyanin and stilbene biosynthesis) was upregulated by water deficit in a tissue-specific manner in the skins of the berries. Other metabolic pathways in the berries were affected by water deficit including ABA, amino acid, carotenoid, lipid, sugar and acid metabolism. Most of these changes were associated with improved quality characteristics of the fruit.

Likewise, Zamboni et al. investigated berry development and withering in grapevine at the transcriptomics, proteomics and metabolomics levels. A multistep hypothesis-free approach from four developmental stages and three withering intervals, with integration achieved using a hierarchical clustering strategy (multivariate O2PLS technique), identified stage-specific functional networks of linked transcripts, proteins and metabolites, providing important insights into the key molecular processes that determine wine quality. A hypothesis-driven approach identified transcript, protein and metabolite variables involved in the molecular events underpinning withering, which predominantly reflected a general stress response. Berry ripening and withering are characterized by the accumulation of secondary metabolites such as acylated anthocyanins, but withering also involves the activation of osmotic and oxidative stress response genes and the production of stilbenes and taxifolin.

Usadel et al. investigated the effects of cold temperatures over time using transcriptomics, metabolomics and enzyme activities. They found some enzyme activities and metabolites changed rapidly, whereas others changed more slowly. The early changes (6 h) in enzyme activities were poorly correlated with transcript abundance, but after 78 h these correlations were greatly improved. Much of the long-term changes in metabolism could be ascribed to the CBF regulon.

Caldana et al. conducted a complex time-series experiment (22 time points) with differing temperatures and light intensities using both metabolomics and transcriptomics analyses. This high-resolution time series experiment revealed that metabolic activities respond more quickly than transcriptional activities, indicating a disconnect between metabolism and transcription in the early phases of stress response and indicating that enzymatic activities may play a significant role. There were common metabolic responses to the changing environment within 1 h of the change including a decrease in energy metabolism and translation and an increase in the transcription of genes involved in signaling cascades. At later time points, condition-dependent metabolism was revealed. For example, protein degradation and energy metabolism derived from amino acids occurred in warm temperatures and darkness. Amino acid catabolism appears to fuel the TCA cycle in the absence of photosynthesis.

Yun et al. characterized the response of rice to a mild chilling stress (10°C). They found that transcriptional regulation consisted of three dynamic and complex phases over 96 h. The early transcriptional phase appeared to be triggered by oxidative signals ($H_2O_2$) and lead to the subsequent induction of cellular defense and rescue mechanisms. Combining temporal co-expression data from microarrays with promoter motif enrichment analyses and oxidative responses, transcriptional regulatory network models for the different response phases were constructed. A bZIP-TGA transcription factor module (as1/ocs/TGA), one of seven transcription factor modules, was the most connected regulatory module in phase one. Each of the transcription factor modules consisted of clusters of transcription factors exhibiting combinatorial control of the chilling regulon. The speed of the response of this network was associated with chilling tolerance. Chilling-resistant genotypes had a much more rapid and pronounced response of this transcriptional regulatory network than chilling-sensitive genotypes. In addition, the transcription factors identified in this study were located within known growth and stress QTLs in the rice genome.

Integration of omics analysis identifies molecular networks functioning in abiotic stress responses.

Integrated omics analyses have markedly increased our understanding of plant responses to various stresses. These analyses are important for comprehensive analyses of abiotic stress responses, especially the final steps of stress signal transduction pathways.

Integrated analyses of the transcriptome and the metabolome successfully demonstrate connections between genes and metabolites, elucidating a wide range of signal output from ABA under dehydration and the DREB1/CBF transcription factors in response to low temperature. Metabolite profiling reveals that ABA accumulates during dehydration, regulating the accumulation of various amino acids and sugars such as glucose and fructose. In particular, the dehydration-inducible accumulation of BCAAs (branch-chain amino acids), saccharopine, proline, and agmatine are correlated with the dehydration-inducible expression of their key biosynthetic genes (BCAT2, LKR/SDH, P5CS1, and ADC2, respectively), which are regulated by endogenous ABA. In addition, metabolome analysis of transgenic Arabidopsis overexpressing DREB1A/CBF3 reveals that there is a striking similarity between the low-temperature regulated metabolome (monosaccharides, disaccharides, oligosaccharides and sugar alcohols) and that regulated by the DREB1A/CBF3 transcription factor. In particular, the low-temperature-inducible accumulation of galactinol and raffinose is correlated with the expression of the Gols3 gene, which is a direct target of DREB1A/CBF3. Maruyama et al. also analyzed DREB2A overexpression, which did not increase the level of any low-temperature regulated metabolites in transgenic plants. Overexpression of DREB2A-CA in transgenic plants increased their tolerance to dehydration stress, but only slightly increased their tolerance to freezing stress. These results indicate that the increased tolerance to freezing stress in transgenic plants overexpressing DREB1A may depend on the accumulation of low-temperature regulated metabolites, especially sucrose, raffinose, galactinol, and myo-inositol. Similarly, transcriptomics and metabolomics analyses of PSEUDO RESPONSE REGULATOR (PRR) arrhythmic triple mutant revealed that the DREB1A/CBF gene and raffinose amounts appear to be regulated by the circadian clock, varying between day and night as if in anticipation of the colder night temperatures.

Comparing metabolomics between dehydration, salinity, light, heat or low temperature stress have identified metabolites that are generally important in abiotic stress responses or are specific to each stress. In a metabolite profiling study of Arabidopsis responses to combined dehydration and heat stresses, heat stress reduced the toxicity of proline, indicating that during the more severe combined stress treatment, sucrose replaces proline in plants as the major osmoprotectant. Comparative metabolite analysis between Arabidopsis responding to heat shock and cold shock revealed that the majority of metabolites in response to heat shock overlapped with those produced in response to cold shock. These results indicate that a metabolic network of compatible solutes includes proline, monosaccharides (glucose and fructose), galactinol, and raffinose, which have an important role in tolerance to temperature stress. Wienkoop et al. identified a RNA-binding protein (ATGRP7) that increased in response to low temperature stress and decreased in response to high temperature stress. Its abundance was significantly correlated with glutamine and proline concentrations. While raffinose and galactinol concentrations were significant markers for temperature responses, their response was independent of the responses of ATGRP7, proline and glutamine.

Transcriptomics, metabolomics and enzyme activities were integrated in a comprehensive study of K deficiency. Carbon and nitrogen metabolism were significantly affected by K deficiency. This integrated approach pinpointed that pyruvate kinase activity (not transcription) was inhibited directly by K deficiency and was primarily responsible for the metabolic disorders observed.

Systematic application of omics technologies has contributed to the development of stress-tolerant crops in the field.

Many genes affect stress tolerance, but few of the identified genes have proven useful in the field. Due to the complexity of stress interactions and stress responses, relevant phenotyping needs to be performed (including field experiments) in abiotic stress studies if we are to make significant progress. The following studies are discussed to highlight good examples of systems biology and omics approaches that have been used to identify key genes regulating stress tolerance and then followed with validation of those responses and phenotypes in multiple experiments including field conditions.

A SNAC1 gene was identified from microarray experiments of stress treatments on rice. SNAC1 is a NAC transcription factor that induces the expression of a number of stress-tolerance genes and improves the drought and salt tolerance of rice in the field. The transgenic plants exhibited increased sensitivity to ABA and reduced water loss. In another drought stress study, a LEA (late embryogenesis abundant) gene was identified from microarray experiments of rice and was transformed and tested in the field under drought conditions through the T3 generation. Spikelet fertility appears to be the main factor contributing to improved yields under drought conditions.

An exhaustive screen of greater than 1500 transcription factors in Arabidopsis identified approximately 40 transcription factors that when overexpressed, improved stress tolerance. One of these transcription factors NF-YB1 was further characterized and shown to display significant drought tolerance in Arabidopsis. Microarray data of this overexpressing line showed few differences in gene expression and the genes identified were not known previously to be involved in drought tolerance. This functional genomics approach provided a new strategy for improving drought tolerance in plants. A homolog of NF-YB1 was cloned in maize (ZmNF-YB2), overexpressed and tested for drought tolerance in the greenhouse and field plots. The transgenic maize lines were more drought tolerant having increased chlorophyll content, photosynthesis, stomatal conductance and grain yields. One line consistently had more than 50% yield improvement in drought conditions over two different years.

Oh et al. used microarrays to identify 42 AP2 transcription factors whose expressions were affected by stress. Two of these transcription factors, AP37 and AP59 were functionally characterized. The two transcription factors are closely related but have distinct differences in affecting rice phenotype. AP37 responded to drought, salinity, cold and ABA; over-expression improved stress tolerance to all three environmental conditions. AP59 responded to drought and salinity, but not cold or ABA, and improved stress tolerance to drought and salinity only. Both overexpressing lines showed improved photosynthetic efficiency under stress conditions. Overexpression of the transcription factors induced common and distinct sets of genes. T5 homozygous overexpressing lines of AP59, but not AP37, had yield penalties under normal paddy conditions in the field, whereas AP37 overexpressing lines, but not AP59, had enhanced yields under drought conditions in the field. The reduced yields of the overexpressing lines of AP59 were attributed to effects on spikelet development.

## Mapping Stress Responses has Provided New Insights and Identified Gaps in our Knowledge of Abiotic Stress Responses

From a meta-analysis of drought-stress related papers from the last 15 years, a complex model for plant responses to drought stress was produced. This model details the interactions of sugars, ROS/RNS, hormones (ABA, ethylene, auxins, cytokinins, salicylic acid, gibberellin and brassinosteroids) and nitrogen metabolism. It highlights the highly complex nature of stress responses.

Much of the signaling involves phosphorylation cascades that react quickly in the plant cell, emphasizing the need for proteomics data as well as transcriptomics data in future models. The PYR/PYL/RCAR-PP2C-SnRK2 pathway illustrates that protein phosphorylation and dephosphorylation are the most important factors in ABA signaling. Similar phosphorylation and dephosphorylation processes are involved in ethylene and other abiotic stress signaling pathways. Not all connections could be drawn in this two-dimensional figure without obscuring many other connections. For example, the interactions of ROS with abiotic stresses and hormones are too complex to display here. In addition, the actual signaling response will be dependent upon the signaling pathway present in that organ, tissue or cell at the time of the response. One needs to use more sophisticated bioinformatics programs like Cytoscape and its plug-ins to visualize the interactions comprehensively in two dimensional or three-dimensional space or with time series views, which would allow these data to be viewed in four dimensions.

Although there are still some technological issues that must be solved to produce a complete picture of protein phosphorylation, several technologies have been developed for the large-scale analysis of phosphoproteins, known as 'phosphoproteomics'. Mass spectrometry analyses have identified thousands of phosphoproteins in Arabidopsis, rice, and Medicago truncatula. In addition, two studies have reported ABA-responsive changes in the phosphoproteome. Phosphoproteomics analyses of mutants for abiotic stress signaling (e.g. PP2C or SnRK) will identify the relevant network of protein phosphorylation events in abiotic stress signaling.

Transcriptome analysis technologies have advanced to the point where high-throughput DNA sequencers and high-density microarrays such as tiling arrays are readily available. These technologies provide new opportunities to analyze noncoding RNAs and can clarify aspects of epigenetic regulation of gene expression. Similar approaches have revealed the global transcriptomes of plants exposed to abiotic stresses such as dehydration, cold, heat, high-salinity, osmotic stress, and ABA. These analyses indicate that these stresses increase or decrease transcript abundance from not only previously identified stress-responsive genes, but also from thousands of unannotated non-protein-coding regions. Matsui et al. estimated that approximately 80% of previously unannotated upregulated transcripts arise from antisense strands of sense transcripts. There was a significant linear correlation between the expression ratios (stress-treated/untreated) of the sense transcripts and the ratios of the antisense transcripts. Interestingly, the data suggested that such stress-responsive antisense transcripts are derived from antisense strands of the stress-responsive genes, RD29A and CYP707A1. Clearly, transcriptional regulation is far more complicated than we previously imagined. Whether or not such antisense transcripts have biological functions is an important issue that remains to be resolved.

Much more research is required in order to fully map plant responses to abiotic stress. The nature of the pathway responses will vary and is highly dependent on the species, organ, tissue, cell type, developmental stage of the plant, the stress or stresses affecting the plant, the level and duration of the stress. Despite the vast amount of research collected on abiotic stress in the last decade, there are still significant gaps in our knowledge.

# ABIOTIC STRESS IN HARVESTED FRUITS AND VEGETABLES

Harvested fruits and vegetables can be potentially exposed to numerous abiotic stresses during production, handling, storage and distribution. Some of these stresses can be minor in nature, resulting in no quality loss or, in some cases, in quality improvement during distribution. However, when the abiotic stress is moderate or severe, quality losses almost always are incurred at market. As a consequence it is important to understand the nature and sources for abiotic stresses that affect fruits and vegetables. In addition, with improved understanding, options for better management or resistance become available. One of the challenges facing fruit and vegetable production globally is that regional climate regimes are becoming more unpredictable from year to year. Hence understanding of effects of field abiotic stresses (e.g. drought, extreme temperatures, light and salinity) on postharvest stress susceptibility will become more important since postharvest stresses limit the storage and shelf life potential of fruits and vegetables.

## Major Preharvest Stresses which Influence Postharvest Abiotic Stress Response

Abiotic stresses occurring during production can either be the primary cause (direct) for disorders that exhibit themselves during postharvest handling and storage practices or they can influence the susceptibility of a fruit or vegetable to postharvest conditions that cause abiotic stresses resulting in disorders (indirect). It is important to characterize the relationship between preharvest abiotic stresses occurring during production and postharvest abiotic stresses that the fruit or vegetable is exposed after harvest and during storage and distribution, since the solution to these different problems will be best resolved by focusing on preharvest or postharvest abiotic stress amelioration, respectively. Moderate levels of preharvest stress can potentially work towards enhancing stress resistance of a fruit or vegetable through up-regulating genes and pathways which renders the tissues cross-tolerant to many stresses which may occur subsequently in postharvest handling, storage and distribution.

## Drought

The occurrence of drought conditions during production of fruit and vegetable crops is becoming more frequent with climate change patterns. While much work has been devoted to understanding of drought effects on production and productivity of these crops, there is limited published literature on the effects of pre-harvest water stress on responses to postharvest stresses and hence on subsequent quality and shelf life.

Water stress during the production phase of some fruits and vegetables may affect their physiology and morphology in such a manner as to influence susceptibility to weight loss in storage. There have been both positive effects reported for field water deficits (stress) in tree fruits and root vegetables. In the case of peaches, it has been shown that lower levels of irrigation results in higher density of fruit surface trichomes and consequent lower weight losses in storage. In addition, two studies have shown that deficit irrigation of apples and pears could reduce water loss of these fruit in subsequent storage and this was attributed to reduction in skin permeance of the deficit irrigated fruit. Presumably, fruit grown under moderate water stresses imposed by deficit irrigation practices adapt by developing a less water permeable cuticle. In terms of understanding that water deficits can have negative effects on postharvest stress susceptibility, irrigation of apples has been shown to enhance apple size which was associated with lower to water losses during storage. This observation highlights a main concern about using deficit irrigation, which is the reduced size of fruit from such treatment. Size of fruit is important, since larger fruit have lower surface area to volume ratios, which confers lower relative water loss. Another negative affect associated with water deficits is the case of root vegetables, such as carrot, where preharvest water stress (watering to 25-75% of soil water field capacity) can weaken the cells, resulting in higher membrane leakage (i.e. cell damage) and consequently greater weight loss in storage.

Timing of a water stress event can also be very important in determining response to postharvest abiotic stress response. One example is that 'Kensington' mango fruit (Mangifera indica L.) will be significantly more susceptible to postharvest chilling injury with exposure to water stress during the cell expansion phase of growth as opposed to being exposed to the stress during cell division or at a time near to harvest maturity. Therefore it is critical to avoid water stress until the fruit has reached maximum size in order to minimize incidence of chilling-induced injury in storage. Water stress, particularly at the tuber forming stage, can also lead to a higher susceptibility of potatoes to postharvest development of black spot disorder. Black spot disorder is correlated primarily to susceptibility of cells in the potatoes to undergo decompartmentation in response to mechanical stress (i.e. bruising).

## Plant Nutrition

Calcium nutrition during production has been well documented in regard to postharvest disorders of many fruits and some vegetables. Calcium is also been suggested as a putative signalling molecule involved in the development of cross-tolerance to abiotic stresses. Therefore the role of preharvest calcium nutrition is postharvest stress resistance may be complex, and dependent on whether the fruit or vegetable is also exposed to environmental abiotic stresses.

Potassium nutrition has been shown to have a few important effects on postharvest abiotic stress susceptibility of vegetables. In carrots, deficiency in potassium is associated with greater weight loss (desiccation stress) in storage. At levels below 1 mM potassium in the soil medium, weight loss was directly associated with increased membrane leakage (i.e. damaged cells) in the carrot tissues. Above 1 mM potassium, there were no significant differences in weight loss under standardized storage conditions. Improved potassium nutrition has also been shown to reduce susceptibility of potatoes to internal bruising in response to mechanical stresses imposed during postharvest handling.

Relatively high preharvest nitrogen is often associated with poor postharvest quality of many fruits and vegetables. In regards to affecting susceptibility to postharvest stress, applying higher than recommended levels of preharvest nitrogen for a specific crop have been linked to storage discoloration susceptibility in both cabbage and potato. In the case of cabbage, it appears that excessive nitrogen fertilization leads to high accumulations of zinc and aluminum and nitrate induced manganese deficiency). High nitrogen applications in the field resulted in increased incidence and severity of black midrib in cold storage, particularly for the susceptible cultivar, 'Safekeeper'. In the case of potatoes, black spot susceptibility (a consequence of bruising; is influenced by nitrogen fertilization, particularly the balance of nitrogen applied in relation to levels of potassium applied. In contrast nitrogen deficiency or lower than recommended nitrogen application rates will most often results in increased vitamin C content in many fruits and vegetables. Vitamin C content has been tightly linked with storage life potential, which is likely a consequence of the importance of this antioxidant nutrient in forestalling oxidative injury that leads to quality losses in storage.

## Temperature Extremes

Susceptibility to high (heat injury inducing) or low (chill injury inducing) temperatures is known to be reduced by prior exposure of the sensitive fruit or vegetable to low ambient temperatures. However, if the preharvest temperature leads to chilling induced injury in the field, then susceptibility to postharvest chilling injury can be increased. Therefore, the level of the preharvest temperature extreme will be a determinant as to if the exposure will have positive or negative effects on postharvest stress sensitivity.

Extreme high temperatures can occur in the field and apple fruit exposed to direct sunlight can reach in excess of 40 C. High temperatures during the late season (leading up to harvest) can enhance susceptibility of apples to superficial scald which develops in storage.

## Salinity

Tomatoes grown under high salinity will produce smaller fruit with higher soluble solids. Smaller fruit will have higher surface area to volume ratios, hence greater susceptibility to postharvest water loss (i.e. desiccation stress). While there is no direct information to confirm that smaller tomato fruit from saline growing conditions would be subject to greater desiccation stress postharvest, firmness declines for tomatoes grown under 3 and 6 dS $m^{-1}$ salinity levels were increased by 50 to 130%, respectively, at two weeks holding at 20 C compared with control fruit.

## Light

It would be considered logical to assume that effects of exposure to high light are difficult to dissociate from effects exposure to high temperatures. However, research has shown that low light (bagging of apples) in the preharvest interval reduced susceptibility of apples to developing superficial scald in cold storage and, in contrast, high ambient temperatures resulted in increased susceptibility. Generally, only sun-exposed surfaces of susceptible cultivars of apples develop scald in storage.

In the case of ambient low light, when lettuce is grown under low light which is sub-optimal for photosynthetic activity, shelf life of fresh cut lettuce (i.e. lettuce subjected to mechanical stress) is much shorter than lettuce produced under optimal light conditions. Tomato size is smaller when the crop is grown under ambient low light levels, such as in the early spring season in northern latitudes and since surface area to volume ratio is greater in smaller fruits, susceptibility to postharvest desiccation stress would increase. Low light also results in lower levels of ascorbate in many greenhouse-grown fruits and vegetables, which would render them less fit to deal with postharvest stresses since ascorbate contents are general directly proportional to relative levels or stress tolerance.

## Postharvest Stresses during Handling and Storage

### Temperature Extremes

Postharvest temperature abuse during distribution is an ongoing challenge for many products, particularly those being shipped by air or ocean container. Breaks in cool chain temperatures can result in acceleration of climacteric ripening and softening (i.e. reduction in shelf potential) for apples harvested and stored at the pre-climacteric stage of maturity. However, those authors also found that temperature breaks had minimal effects on apples harvested and stored at post-climacteric stage of maturity. Fruit harvested at post-climacteric stages of maturity are not generally of concern to industry since those fruit have shorter storage and shelf life potential.

Chilling injury susceptibility is a significant issue for many crops derived from subtropical and tropical growing regions. Generally fruits, fruit vegetables (fruits which are consumed as vegetables) and root and tuber crops are chilling sensitive. There is a significant literature pertaining to chilling sensitivity of crops and much effort has been devoted to better understanding of chilling injury mechanisms and approaches to ameliorating the disorder. While there are visible (surface pitting, internal browning) and textural (accelerated softening and development of mealiness) changes often associated with chilling injury, flavour generation capacity has been shown to be a sensitive early indicator of chilling stress effects.

Use of heat treatments has become popular for disinfestation and disinfection of fruits and vegetables. Appropriately applied levels of heat treatment can also induce temperature tolerance to both high and low temperatures. However, when excessively harsh heat treatments are applied, heat-induced damage can occur.

## Low $O_2$ and High $CO_2$

Managing postharvest handling and storage atmospheres to avoid low $O_2$ or high $CO_2$ stress is a constant concern and the problem is more severe when handling products in modified atmosphere (MA) packages as opposed to controlled atmosphere (CA) systems since temperature is often not as easily controlled in the MA packages as the produce moves through a distribution chain. As such, the importance atmospheric stress in postharvest systems deserves significant discussion.

While low $O_2$ levels are well-known to induce stress-induced changes in metabolism and resultant metabolite accumulations, acute low oxygen injury is not expressed until the tissue is re-aerated and a consequent uncontrolled oxygen burst (consisting of hydrogen peroxide and other radicals) occurs, resulting in lipid peroxidation, protein denaturation and membrane injury. Different fruits and vegetables have varying thresholds for low $O_2$ stress, dependent on anatomy, temperature, physiological age, presence of supplemental gases (e.g. $CO_2$, CO, $SO_2$, $C_2H_4$) and duration of exposure. Threshold limits for low $O_2$ tolerance can range from approximately 10 kPa partial pressure (for early potatoes and asparagus) down to approximately 1 kPa partial pressure (for mushroom, broccoli, and chipping potatoes), assuming that the produce is cooled to its recommended temperature when it is placed into that atmosphere. When a fruit or vegetable is converted to fresh-cut format, it generally becomes tolerant to lower levels of oxygen.

One of the most notable effects of high $CO_2$ levels in postharvest handling is to competitively inhibit ethylene binding and action hence delaying ripening in climacteric fruits. High $CO_2$ will directly inhibit succinate dehydrogenase, thus impairing the functioning of the tricarboxylic acid cycle and aerobic respiration. There are numerous physiological disorders that can be attributed to high $CO_2$ stress, including black heart of potatoes, brown heart or core in apples and pears, surface bronzing in apples and brown stain of lettuce. High $CO_2$ can also modulate chilling stress, ethylene induced disorders and susceptibility to pathogenic attack.

## Mechanical Injury

There are two types of mechanical injury that can be incurred during harvest and handling of fruits and vegetables; (1) cuts or punctures, and (2) impacts leading to bruises. Cuts can lead to transitory increases in respiration, ethylene production, phenolics production and cell deterioration near the site of the injury. Several factors influence the severity and size of bruising sustained, including maturity, water potential, tissue or cellular orientation at the site of the injury, shape of the object imparting the bruising force, energy and angle of the impact, and temperature of the product.

Cut type injuries are most prevalent in fresh-cut fruit and vegetable products. Severity of response to cutting is very much dependant on the tissue characteristics, maturity of the fruit or vegetable of interest, the coarseness or sharpness of the cutting implement used and the temperature at which the cutting is done. Cut injuries occur during the harvest process of many fruits and vegetables and is more severe in machine-harvested as compared with hand-harvested product. Products such as tomatoes, pickling cucumber, snap beans, green peas, potatoes, garlic can sustain significant damage during machine harvest, whereas asparagus, Brussels sprouts, leafy greens and head lettuce are more tolerant. Cutting associated disorders are primarily visualized as cut-edge browning or blackening, but may also include yellowing in green tissues and whitening on carrots.

Impact caused injuries are associated with loading of product for transport, events during transport (particularly when uneven or rough roads or lanes are encountered), during unloading and throughout packaging and processing lines. Impact bruising has been shown to induce significant levels of ethylene production in mature-green tomatoes, with the levels produced being concomitant with the number of impacts sustained. Both physically visible internal damage and slight declines in ascorbic acid content of the fruit were observed in response to the impact events imposed on the fruit. Impact bruising leads to development of internal black spots in potato tubers.

## Desiccation

Loss of water leading to deterioration in fruit and vegetable tissues is a common issue for postharvest handling and distribution. In addition to wilting, water stress can lead to accelerated senescence or ripening which are expressed as softening of tissues, membrane deterioration and yellowing. As mentioned previously, one of the main characteristics of a fruit or vegetable that defines susceptibility to water loss is surface area to volume ratio. Burton described a range of surface area to volume ratios from as low as 0.2 $cm^{-1}$ in winter cabbage or turnip to as high as 50-100 $cm^{-1}$ in leafy vegetables, explaining differences in water loss characteristics in these very different types of vegetables. In beans, the density of hairs on the cuticle can modulate rates of water loss to some extent and damage to these hairs will lead to increased losses. In other fruits and vegetables, lenticel density or epicuticular wax thickness can modulate water loss.

The driving force for water loss is the vapour pressure deficit (vpd), which is the relationship that describes the difference in water activity of the fruit or vegetable and the water activity of the atmosphere surrounding it. The greater the vapour pressure deficit, the greater the water loss. Three postharvest handling principles are important in minimizing water loss of any fruit or vegetable; (1) warm product loses water faster than cool product when placed into a cool room, hence the importance of rapid precooling before storage, (2) delays in cooling will lead to longer exposures to higher vapour pressure deficit conditions, hence timely cooling after harvest is of utmost importance, and (3) storing product at the coldest storage temperature and highest relative humidity possible will minimize water loss. Following these principles will result in the least desiccation stress for any fruit or vegetable.

## Effects of Abiotic Stresses on Plant Metabolism

### Metabolic Changes Induced by Stress

Heat stress induces metabolic changes associated with heat shock protein accumulations which are known to confer persistent levels of stress resistance in heat-exposed produce. Heat stress can also inhibit the production and accumulation of lycopene. The duration and temperature of exposure will determine if such an effect will occur, but tomato fruit exposed to 32°C continuously will not turn red, remaining yellow even at full ripeness.

Chilling stress results in the accumulation of lipid peroxidation products, superoxide anions and hydrogen peroxide and also to losses in flavor volatile production in sensitive fruits and vegetables. Chillingstress affects ethylene metabolism, leading to accelerated softening. It also modifies cell wall metabolism, with resultant up-regulation of cell wall dismantling enzymes such as pectin methylesterase and endopolygalacturonase.

There are two distinct phases for response to exposure to anaerobic or anoxic atmospheres. The first phase involves the metabolic shift induced by impairment of the mitochondrial electron transport chain by limitation of the primary electron acceptor, molecular $O_2$. The resultant impairment has been well-characterized as declines in ATP levels, pyruvate dehydroxylase activity and cytoplasmic pH. These conditions lead to an increase in pyruvate decarboxylase, alcohol dehydrogenase and lactate dehydrogenase activities. Anaerobic respiration is induced and there is an accumulation in acetaldehyde, ethanol, ethyl acetate and lactate. In addition, the impairment of the mitochondrial electron transport chain results in electron leakage and consequent generation of superoxide anions and hydrogen peroxide in the cells which can be decomposed by existing cellular antioxidant systems. Ascorbate and glutathione levels can also increase during an anoxic or hypoxic event. Any or all of these changes are indicators of oxygen stress in fruit and vegetable tissues and affect quality attributes. However, actual injury to the tissue does not occur until the second phase, where the fruit or vegetable is returned to higher $O_2$ atmospheres. On return to aerobic atmospheres, rates of oxygen radical production in the impaired electron transport chain becomes accelerated, leading to great accumulations of superoxide anion, hydrogen peroxide and hydroxyl radical that cannot be fully decomposed by existing antioxidant protection systems and consequent lipid peroxidation and protein denaturation ensues, leading to membrane damage, enzymatic browning and cell death.

Kubo et al. found that different fruits and vegetables responded in different ways to high $CO_2$. Those fruits and vegetables showing no change in $O_2$ uptake (respiration) in response to 60% $CO_2$ exposure did not show any change in ethylene production. Those showing a decline in respiration generally also showed a decline in ethylene production, except banana and those which showed a significant increase in respiration also showed an increase in ethylene production (cucumber, eggplant, podded pea, spinach and iceberg lettuce). This suggests that these latter vegetables were more sensitive to $CO_2$ stress than the other fruits and vegetables tested. Mathooko et al. determined that $CO_2$ stress-induced ethylene biosynthesis was likely regulated at a post-transcriptional level and it involves the de novo synthesis of novel proteins. They also identified that protein phosphorylation and dephosphorylation processes as being required in one or more of the steps leading to the induction of ACC synthase, which is the last enzyme in the ethylene biosynthesis pathway.

Initial response to wounding stress is characterized with a progressive accumulation of ACC synthase, ACC and ethylene production in tomatoes and this lasts for up to 2 hours, but if the cut tomatoes are held over longer durations, ethylene production declines. However, both ACC synthase and ACC continue to accumulate, indicating that there is a capacity to produce ethylene, but transformation of ACC to ethylene is blocked or inhibited. Wounding has also been shown to induce phenolic accumulation through up-regulation of phenylalanine ammonia lyase (PAL). This up-regulation of PAL was associated with wound-induced ethylene production. Other metabolites increase in response to bruising or wounding stress, including isocoumarin in carrots, anthocyanins in midribs of red-pigmented lettuce, methanethiol, allyl isothiocyanates and dimethyl disulfide in cabbage, six-carbon aldehydes and alcohols in cut peppers and suberin in tomatoes and bean pods. Some of the metabolic shifts are mediated by stress response messengers (e.g. phenolics, suberin and isocoumarin accumulation) and others are a direct consequence of cellular disruption that occurs during wounding or bruising (e.g. methanethiol, allyl isothiocyanate and dimethyl sulphide accumulations).

Desiccation stress has been shown to cause some metabolic changes in fruits and vegetables. In carrots, desiccation stress at extreme handling conditions (i.e. at 13°C) led to increases in osmotic potential, which are a function of free sugars in the roots. Increase in osmotic potential in response to water loss was most likely explained by increases in polysaccharide hydrolyzing enzymes activities in response to the stress. Hence, enzymes such as polygalacturonase and pectinesterase may increase in activity leading to loss of cell wall structure and concomitant increases in soluble sugars. This may explain at least a component of the loss of firmness that has been observed with carrots as they lose water. This hypothesis is borne out by results of work with cucumbers where water stress resulted in up-regulation of polygalacturonase activity, suggesting that water loss itself was not the only factor in causing softening of stressed fruit. Another aspect of water stress is induction ethylene production, which may explain why water stress leads to accelerated ripening in bananas and accelerated senescence in bell peppers.

## Effects on Quality

Postharvest abiotic stressors can lead to numerous quality problems in fruits in vegetables, including scald, core and flesh browning of fruits, sweetening, pitting, water-soaking appearance, abnormal ripening, russeting and tissue softening. Stress can also result in the losses of nutrient constituents in the fruit or vegetable, with vitamin C loss being the most sensitive indicator of stress exposure.

## Mechanisms for Abiotic Stress Response at the Biochemical and Molecular Levels

Response to all abiotic stresses can be acute and sub-acute in nature, where acute responses represent cases where cell death is a direct result and sub-acute responses represent cases where the stress leads to induction of adaptive changes in biochemical and gene expression. Many reactive oxygen species (ROS), particularly hydrogen peroxide, behave as signalling agents to trigger biochemical changes at the gene expression level. In general, abiotic stressors will induce perturbations in the fruit or vegetable cellular homeostasis which will then result in the increased generation of ROS in the apoplast, mitochondria, peroxisomes, cytoplasm, chloroplasts and endoplasmic recticulum. The ability of the cell to initially cope will depend largely on the endogenous free radical scavenging capacity.

When free radical generation exceeds the endogenous scavenging capacity, the ROS interact with sensors, for which the full nature are not currently understood, that will initiate mitogen activated-protein kinase (MAPK) cascade reactions and also directly up-regulate transcription factors and calcium/calmodulin kinases. The MAPK cascade reaction will activate various transcription factors that enable de novo production of ROS, ROS scavenging systems, accumulation of heat shock proteins, and modulate NADPH supply in the cell. Some of the MAPK cascade paths have been also shown to be linked specifically to ethylene production, which is probably why ethylene production seems to be intrinsic to most stress responses. However, not all stressors produce identical response pathways, and so there is still a lot of work to be done in mapping of stress response networks.

Chilling, bruising and cutting injuries all lead to increased activities of cell wall hydrolysing enzymes (polygalacturonase and pectinmethylesterase), which accounts for accelerated softening and abnormal ripening that occurs in response to those abiotic stresses. Water stress will also lead to accelerated softening, and that response has been associated with the induction of ethylene production in response to water stress. It quite likely that the effect of all these stresses on up-regulating the cell wall hydrolysing enzymes and consequently accelerating softening is an ethylene-mediated response.

Accumulation of heat shock proteins (HSPs) which are mediated via transcription factor activation downstream of the MAPK cascade have been reported to enhance persistent levels of stress resistance in affected tissues. HSPs are thought to be an important factor for protein folding, assembly, translocation and degradation under normal, stress free conditions. HSPs have also been linked to stabilization of proteins and membranes, and enabling protein refolding under stress conditions. As a consequence HSPs are thought to have a pivotal function of protecting plant tissues against stress by maintaining cellular homeostasis.

## Approaches to Ameliorate Abiotic Stress Sensitivity

### Treatments to Enhance Stress Resistance

Numerous postharvest treatments have been evaluated for enhancing abiotic stress resistance of fruits and vegetables. Generally, temperature modulation (including intermittent warming), extreme atmospheres (high $O_2$, $CO_2$ and low $O_2$), growth regulators, anti-transpirants, antioxidant dips, growth regulators, nitric oxide and ethanol have been tested. Application of treatments in combination can often improve the stress resistance level to the fruit or vegetable.

One treatment that has not been widely studied to date is the concept of gradual cooling to enhance resistance to chilling injury. Gradual cooling (2°C per day) has been shown to reduce susceptibility of tomato to chilling injury, presumably by cooling slowly enough to allow the inherent stress resistance systems to develop before actual chilling temperatures were reached. The approach is perhaps more amenable to commercial practice than some others tested (i.e. intermittent warming).

While there has been significant effort placed on developing treatments to enhance postharvest abiotic stress resistance in fruits and vegetables, the efforts have only resulted in incremental success. This is largely due to the fact that stress response and resistance is a very complex matrix of processes and pathways, which are not fully understood at this point in time. Hence, it is very difficult to design effective treatments to achieve the resistance required for any or all abiotic stresses that a fruit or vegetable may encounter during harvest, handling, storage and distribution.

Hot or warm water treatments have been shown to minimize cutting induced injuries in fresh-cut products. Such treatments may also be used to control chilling injury via induction of heat shock proteins. Warm water treatment has also been shown to reduce sensitivity to irradiation in cut and packaged lettuce. Each product has a differing tolerance to temperature and so such treatments must be developed case by case for each vegetable or fruit of interest. The key is apply a sublethal temperature exposure, in order to enhance the adaptive response of the fruit or vegetable tissue.

Another approach is to apply conditioning treatments to enhance chill stress tolerance. This has been shown to provide enhanced chilling stress in several fruits and vegetables. It also has the advantage over heat treatments in that the risk of causing acute injury during treatment is very low.

Atmospheric treatments are quite often beneficial to controlling response to postharvest stresses. Modified or controlled atmospheres have been shown to help minimized chilling injury in a number of fruits and vegetables. However, it must be noted that in some cases, application of a controlled atmosphere led to detrimental effects on the fruit (e.g. asparagus, cucumber, limes, sweet bell peppers). Therefore, a case by case evaluation is required to determine if secondary stress is induced by the atmosphere treatments.

Growth regulator applications can also potentially enhance stress resistance, particularly for fruits and vegetables which are prone to show accelerated ripening or senescence in response to stress. Accelerated ripening or senescence is most often mediated by ethylene production in response to the stress. As consequence anti-ethylene products such as aminovinylglycine (AVG) and 1-methylcyclopropene (1-MCP) could enhance storage or shelf if ethylene production in response to stress was a main concern. Other growth hormones, such as methyl jasmonate (which promotes leaf senescence), can enhance chilling resistance in avocado, grapefruit, bell peppers and zucchini squash. Abscisic acid has been demonstrated to reduce chilling-induced injury in some crops. Other growth regulators have been suggested for use in preventing senescence in leafy vegetables (e.g. 2, 4-D), however their practical application is limited.

Quite often a single stress resistance enhancing treatment may not confer optimal resistance levels to all postharvest stresses encountered. Therefore it may be productive to consider application of a combination of two or more stress tolerance enhancing treatments to achieve optimal levels of resistance. There are a several examples which show the added benefit of applying treatment combinations.

## Germplasm Selection

There have been some reports on the selection of germplasm and cultivars from breeding programs that will have greater postharvest stress resistance and hence better storage capability. Hodges et al. were able to show that a cultivar which was more susceptible to yellowing was so because of differences in balance of antioxidant systems in the tissues, which resulted in higher accumulations of ROS, particularly hydrogen peroxide. They hypothesized that the higher levels of ROS were directly causal of the yellowing of the chlorophyll in the spinach leaves. A similar relationship was found for a broccoli cultivar resistant to yellowing when compared with a cultivar highly susceptible to yellowing during storage. Apple cultivars which have resistance to developing browning in response to cutting have greater levels of apoplastic antioxidant enzymes and lower levels of ROS in the apoplast after cutting than cultivars with lower levels of antioxidant enzymes. While measures of antioxidant enzymes and ROS levels can be instructive to understand the basis for difference in stress tolerance in selections of a breeding population, they are too labour intensive to consider for incorporation as selection tools into a breeding program.

In vitro selection is an approach where plant cells of a fruit or vegetable of interest are tissue-cultured and exposed to a stressor, taking surviving cells to regenerate new plants having a superior level of stress resistance. This approach has been highly successful to regenerate germplasm of many crop plants which can be regenerated via tissue culture techniques. In vitro selection is much less expensive approach than molecular engineering and laboratories can easily be set up almost anywhere in the world having access to basic utilities, using inexpensive technology. The fact that the approach relies on the cells to survive in response to the applied stress means that the complexity of the stress response network is incorporated into the plants regenerated in this way.

## Use of Molecular Probes for Marker - Assisted Breeding

There are a large number of genes and proteins associated with stress tolerance in plants and so the best approach to identifying stress tolerant lines is apply the stress of interest and perform quantitative trait loci (QTL) analyses. This approach may be used with intact plants and harvested plant parts, generally using the plant part which is of interest in the breeding improvement strategy. The approach requires analysis of adaptive changes in QTL expression as opposed to constitutive expression. A stress protocol, to which target fruit or vegetable will be exposed, must be established to provide clear differentiation between resistant and susceptible lines. However, as stated earlier, the stress response is complex and so success using QTLs will require interdisciplinary effort, integrating biochemistry, gene mapping and phenotyping activities to allow reliable interpretation and successful use of adaptive QTLs for selecting for stress resistance.

## Molecular Engineering

Molecular engineering for stress resistance in fruits and vegetables is limited due two major factors; the complexity of the stress response network means that modulating the stress resistance with single gene insertions is unlikely, and methods to successfully transform many important fruit and vegetable crops have yet to be developed. One area where there has been some advance is insertion of anti-freeze genes to protect against low temperature injury. However, the future advances will require a better basic knowledge of the stress response network and control points at the molecular level. This approach may be a helpful adjunct to adaptive QTL analysis, since insertion of a putatively important gene may alter the adaptive QTL profile and provide a probe to better understand the functional changes induced by a stress.

## Improved Postharvest Harvest and Handling Protocols

Most existing postharvest handling and storage procedures are generally not considered to be stress-inducing, except in cases where quality issues arise. However, there is increasing information that suggests that there are significant stresses that could be modulated during the postharvest continuum. It is often difficult to avoid low temperature stress, since most produce is refrigerated as a necessary step to control spoilage and preserve food safety.

Simple modifications to postharvest handling systems can sometimes result in significant reduction in stress exposure and consequently result in storage and shelf life extension. One of the most successful strategies is the application of plastic film packaging or wraps to prevent desiccation, resulting in significant improvements and shelf life and quality of many fruits and vegetables. In many cases, modified atmosphere packaging is considered to largely control humidity around product and thus prevent moisture loss of fresh-cut and whole fruits and vegetables. Also, anti-transpiration coatings have been shown to be effective for maintaining quality through control of water loss.

In regards to maintaining water content on the retail shelf, the application of misting systems can 'recharge' the vegetable and in so doing maintain quality over longer durations at less than ideal storage temperature.

While rapid cooling is generally recommended to preserve quality, delayed or gradual cooling may be useful to chilling sensitive crops to allow them to acclimatize or adapt to storage and handling conditions. Delays in cooling of apples after harvest can be used to reduce the development of low temperature stress induced orders in 'Honeycrisp' apples. Similarly, delays in cooling can reduce $CO_2$-induced internal browning in 'Braeburn' apples. Recently, slow cooling at a rate of 2 C per day from 12 °C to 4 °C of tomatoes has shown promise to reduce chilling injury when stored at the lower temperature.

## PLANT GENES FOR ABIOTIC STRESS

Abiotic stress is the primary cause of crop loss worldwide, reducing average yields for most major crop plants by more than 50%. Plants as sessile organisms are constantly exposed to changes in environmental conditions. When these changes are rapid and extreme, plants generally perceive them as stresses. However stresses are not necessarily a problem for plants because they have evolved effective mechanisms to avoid or reduce the possible damages.

The response to changes in environment can be rapid, depending on the type of stress and can involve either adaptation mechanisms, which allow them to survive the adverse conditions, or specific growth habitus to avoid stress conditions. In fact, plants can perceive abiotic stresses and elicit appropriate responses with altered metabolism, growth and development. The regulatory circuits include stress sensors, signalling pathways comprising a network of protein-protein interactions, transcription factors and promoters, and finally the output proteins or metabolites.

A number of abiotic stresses such as extreme temperatures, high light intensity, osmotic stresses, heavy metals and a number of herbicides and toxins lead to over production of reactive oxygen species (ROS) including $H_2O_2$ causing extensive cellular damage and inhibition of photosynthesis.

Normally, ROS are rapidly removed by antioxidative mechanisms, but this removal can be impaired by stresses themselves, causing a rise in their intracellular concentration and an increase of the damage. To prevent or repair these damages, plant cells use a complex defence system, involving a number of antioxidative stress-related defence genes that, in turn, induce changes in the biochemical plant machinery. Studies have shown that ROS probably require additional molecules to transduce and amplify defence signals. ROS production and anti-oxidant processes, all act in a synergistic, additive or antagonistic way, related to the control of oxidative stress.

Responses to stress are not linear pathways, but are complex integrated circuits involving multiple pathways and in specific cellular compartments, tissues, and the interaction of additional cofactors and signalling molecules to coordinate a specified response to a given stimulus. Onset of a stress triggers some (mostly unknown) initial sensors, which then activate cytoplasmic $Ca^{2+}$ and protein signalling pathways, leading to stress-responsive gene expression and physiological changes.

Table: Consequences of abiotic stress and plant responses.

| Stress | Consequences | Plant Responses |
|---|---|---|
| Heat stress | High temperature lead to high evaporation and water deficit. The consequent increased turnover of enzymes leads to plant death. | Efficient protein repair systems and general protein stability support survival, temperature can lead to acclimation. |
| Chilling and cold stress | Biochemical reactions proceed at slower rate, photosyntesis proceeds, carbon dioxide fixation lags, leading to oxigen radical damage. Indeed, freezing lead to ice crystal formation that can distrupt cells membranes. | Cessation of growth in adaptable species may be overcome by changes in metabolism. Ice crystal formation can be prevent by osmolyte accumulation and synthesis of hydrophilic proteins. |
| Drought | Inability to water transport to leaves leads to photosyntesis declines. | Leaf rolling and other morphological adaptations. Stoma closure reduces evaporative transpiration induced by ABA. Accumulation of metabolities, consequently lower internal water potential and water attracting. |
| Flooding and submergence | Generates anoxic or microaerobic conditions interfering with mitochondrial respiration. | Development of cavities mostly in the roots that facilitate the exchange of oxigen and ethylene between shoot and root (aerenchyma). |
| Heavy metal accumulation and metal stress | In excess, detoxification reactions may be insufficient or storage capacity may exceeded. | Excess of metal ions may be countered by export or vacuolar deposition but metal ions may also generate oxygen radicals. |
| High light stress | Excess light can lead to increased production of highly reactive intermediates and by-products that can potentially cause photo-oxidative damage and inhibit photosynthesis. | Exposure of a plant to light exceeding what is utilized in photochemistry leads to inactivation of photosynthetic functions and the production of reactive oxygen species (ROS). The effects of these ROS can be the oxidation of lipids, proteins, and enzymes necessary for the proper functioning of the chloroplast and the cell as a whole. |

Also accumulation of abscisic acid (ABA) plays an important role in abiotic stress signalling and transduction pathways, mediating many responses.

It is well known that abiotic stresses in general, through regulation of both gene expression and protein turnover, alter the abundance of many transcripts and proteins, indicating that transcriptional and post-transcriptional regulation play an essential role in the adaptation of cellular functions to the environmental changes.

Recent advances in molecular biology, genomics, proteomics and metabolomics have provided insight into plant gene regulatory network system, which is mainly composed of inducible-genes (environmental factors and developmental cues), expression programming and regulatory elements (cis-element and trans-element), corresponding biochemical pathways and diverse signal factors. Genetic studies revealed that stress tolerance traits are mainly quantitative trait loci (QTLs), which make genetic selection of traits difficult.

Responses to abiotic stress require the production of important metabolic proteins such as those involved in synthesis of osmoprotectants and of regulatory proteins operating in the signal transduction pathways, such as kinases or transcriptional factors (TFs). In addition, new transcripts are made and within a few hours a steady level of stress adaptation has been reached. In general, the transcriptional regulation of genes is directly controlled by a network of TFs and transcription factor binding sites (TFBS). TFs are proteins with a DNA domain that binds to the cis-acting elements present in the promoter of a target gene. They induce (activators) or repress (repressors) the activity of the RNA polymerase, thus regulating gene expression. TFs can be grouped into families according to their DNA-binding domain. The presence or absence of transcription factors, activators and suppressors regulating transcription of target genes often involves a whole cascade of signalling events determined by tissue type, developmental stage or environmental condition.

Environmental stress-inducible genes can be mainly divided into two groups in terms of their protein products: one type of genes, whose coding products directly confer to plant cells the resistence to environmental stress such as late embryogenesis abundant (LEA) protein, anti-freezing protein, osmotic regulatory protein, enzymes for synthesizing betaine, proline and other osmoregulators; the other groups of genes, whose coding products play an important role in regulating gene expression and signal transduction such as the transcriptional elements. At least four different regulons can be identified, two ABA independent (1 and 2) and two ABA dependent (3 and 4): (1) the CBF/DREB regulon; (2) the NAC (NAM, ATAF and CUC) and ZF-HD (zinc-finger homeodomain) regulon; (3) the AREB/ABF (ABA-responsive element-binding protein/ ABA-binding factor) regulon; and (4) the MYC (myelocytomatosis oncogene)/MYB (myeloblastosis oncogene) regulon.

Our knowledge of the molecular mechanisms underlying the responses of plants to such environmental stresses is still rather limited, but an increasing number of genes have been identified in recent years that mediate these responses. Some of these genes are induced by stress stimuli and encode products that confer tolerance to adverse conditions, whereas others encode upstream regulators that function within signalling pathways controlling the stress response.

## Abiotic Stress-inducible Genes

The complex plant response to abiotic stress involves many genes and biochemical-molecular mechanisms. The analyze of the functions of stress-inducible genes is an important tool not only to understand the molecular mechanisms of stress tolerance and the responses of higher plants, but also to improve the stress tolerance of crops by gene manipulation. Hundreds of genes are thought to be involved in abiotic stress responses.

Many drought-inducible genes are also induced by salt stress and cold, which suggests the existence of similar mechanisms of stress responses.

These genes are classified into three major groups: (1) those that encode products that directly protect plant cells against stresses such as heat stress proteins (HSPs) or chaperones, LEA proteins, osmoprotectants, antifreeze proteins, detoxification enzymes and free-radical scavengers; (2) those that are involved in signalling cascades and in transcriptional control, such as Mitogen-activated protein kinase (MAPK), Calcium-dependent protein kinase (CDPK) and SOS kinase, phospholipases and transcriptional factors; (3) those that are involved in water and ion uptake and transport such as aquaporins and ion transporters.

## Transcriptional Factor Genes Involved in Abiotic Stress

Plant growth and productivity are under constant threat from environmental changes in the form of various stress factors. The most common abiotic stresses are drought, flooding or submergence, salinity, extreme temperatures (heat and freezing) and high light. Furthermore, the continued modification of the atmosphere by human activities lead to increase in the concentration of ozone in the troposphere and this can generate oxidative stress, which leads to the destruction of proteins and cells, premature ageing and reduced crop yields.

Tolerance or susceptibility to these abiotic stresses is a very complex phenomenon, both because stress may occur at multiple stages of plant development and more than one stress simultaneously affects the plant. Therefore, the perception of abiotic stresses and signal transduction to switch on adaptive responses are critical steps in determining the survival and reproduction of plants exposed to adverse environments.

During the past few years, transcriptome analysis has indicated that distinct environmental stresses induce similar responses. Overlap between stress responses can explain the phenomenon known as cross-tolerance, a capability to limit collateral damage inflicted by other stresses accompanying the primary stress.

Responses to abiotic stresses require the production of important metabolic proteins such as those involved in synthesis of osmoprotectants and regulatory proteins operating in signal transduction pathways, that are kinases or transcription factors (TFs). The regulation of these responses requires proteins operating in the signal transduction pathways, such as transcriptional factors, which regulate gene expression by binding to specific DNA sequences in the promoters of respective target genes. This type of transcriptional regulatory system is called regulon. At least four different regulons that are active in response to abiotic stresses have been identified. Dehydration-responsive element binding protein 1 (DREB1)/C-repeat binding factor (CBF) and DREB2 regulons function in abscisic acid (ABA)-independent gene expression, whereas the ABA-responsive element (ABRE) binding protein (AREB)/ABRE binding factor (ABF) regulon functions in ABA-dependent gene expression. In addition to these major pathways, other regulons, including the NAC (or NAM, No Apical Meristem) and Myeloblastosis-Myelocytomatosis (MYB/MYC) regulons, are involved in abiotic stress-responsive gene expression. Particularly, NAC-type TF OsNAC6 is induced by abiotic stresses, including cold, drought and high salinity. Microarray analysis showed that many abiotic inducible genes were up regulated in rice plants over-expressing OsNAC6.

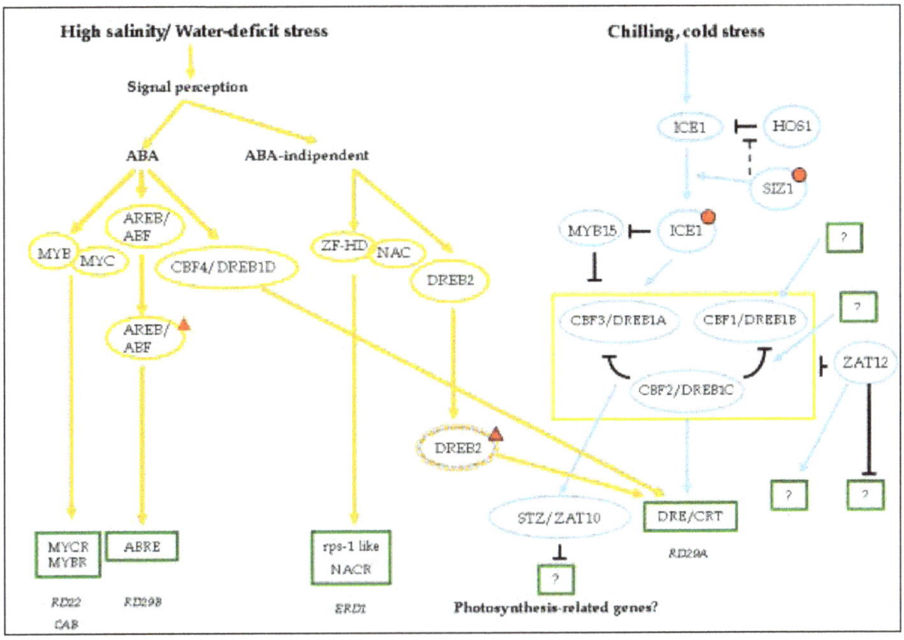


Transcriptional network of abiotic stress responses.

TFs are powerful targets for genetic engineering in abiotic stress resistance in crop plants.

Transcription factors are shown in ovals. Transcription factor-modifying enzymes are shown in circles. The small triangles correspond to post-translational modifications. Green squares with question marks represent putative MYC ICE1-like transcription factors that may activate CBF1/DREB1B and CBF2/DREB1C. The green boxes represent the cis-elements present in stress-responsive genes. The red dot corresponds to the sumoylation modification by SIZ1 of the ICE1 transcription factor. The dashed black line from SIZ1 to HOS1 represents competition for binding places on the ICE1 transcription factor. SIZ1 blocks the access of HOS1 to the ubiquitination sites on the ICE1. CBF4/DREB1D is a DRE cis-element binding factor that is ABA dependent.

## Drought Stress Transcriptional Factors

The genome controls the regulation of the response to water deficit as well as the effectiveness of the response. Microarrays, largely performed using Arabidopsis thaliana as model plant, have been used to catalogue the many genes that are induced or repressed in response to conditions that may lead to cellular water-deficit stress. These genes can be placed in at least four different functional groups: signal transduction, transcriptional regulation, cellular metabolism and transport and protection of cellular structures.

There are at least six different classes of TFs that participate in gene induction or repression in response to water deficit. Homeobox domain and NAC domain containing TFs are induced by multiple treatments that mimic water-deficit stress. Accumulation of proteins which have metabolic or structural functions promote adaptation to stress. One class of genes that could play a role in protection is called the late embryogenesis abundant (Lea) genes. The Lea genes are also developmentally programmed for expression in desiccating seeds. These genes encode small hydrophilic proteins that are predicted to protect proteins and membranes through chaperone-like functions. These proteins were thought to improve the performance of rice plants by protecting cell membranes from injury under abiotic stress.

## Gene Regulation and Transcriptional Factors in Water Deficit

A recent review on analysis of gene expression during drought stress response in plants show and summarize the functions of some genes in both stress response and tolerance. Microarray analysis performed on wheat genome, showed that among 300 unique single expressed sequences tag (ESTs), the 30% of genes were significantly up-regulated and the 18% were down-regulated under drought stress.

Potential functions of approximatively 130 genes of A. thaliana up-regulated in water-deficit was reported by Bray. These genes are involved in cellular response to drought stress by signalling events, detoxification and other functions. cDNA microarray analysison 7000 Arabidopsis full-length cDNAs clarify relationship between rehydratation-, proline and water-treatment inducible genes. Among the 152 rehydratation-inducible genes, 58 genes conteined in their promoter regions the ACTCAT sequence involved in proline and hypoosmolarity inducible gene expression, suggesting that this motif is a major cis-activing element involved in rehydratation-inducible gene expression.

Moreover, microarray analysis performed on two moderately drought-tolerant native Andeon potato clones revealed that there was 1713 differentially expressed genes with 186 up-regulated involved in drought tolerance by inducing of osmotic adjustment, changes in carbohydrate metabolism, membrane modifications and cell rescue mechanisms, such as detoxification of oxygen radicals and protein stabilization.

These recent study underline how the expression of genes in response to water deficit is complex and can be regulated at the transcriptional, post-transcriptional and translational levels. Two major transcriptional regulatory pathways of gene expression play an important role in response to water-deficit stress: the ABA-independent pathway and ABA-dependent pathway. The first is controlled largely by a family of TFs called dehydration response element binding protein (DREB), which contains a DNA binding motif originally identified in a flower patterning protein called APETALA2 (AP2), while transcription factor families known to be as the most responsive to ABA signalling under drought are NAC, AREB/ABF, and MYB.

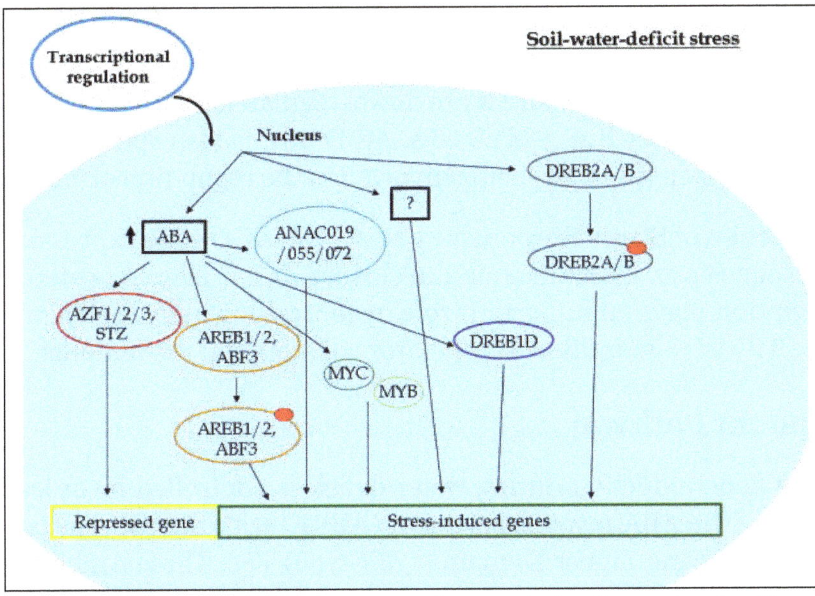


The inset shows the different types of transcription factors involved in induction/repression of regulons. Classes of genes that are induced by water-deficit stress.

## ABA-independent Pathway

DREB are important TFs which induce a set of abiotic stress-related genes and confer stress resistance to plants. The DREB TFs could be divided into two group: DREB1, involved signal transduction pathways under low temperature; DREB2, involved in signal transduction pathways under dehydration. They belong to the ethylene responsive element binding factors (ERF) family of TFs. ERF proteins are a sub-family of the AP2/ethylene responsive element binding protein (EREBP) TFs that is distinctive to plants. ERF proteins share a conserved 58–59 amino acid domain (the ERF domain) that binds to cis-elements, the GCC box, found in many pathogens related (PR) gene promoters conferring ethylene responsiveness, and to the C-repeat CRT/dehydration responsive element (DRE) motif involved in the expression of cold and dehydration responsive genes.

The DREB proteins contain an ERF/AP2DNA-binding domain quite conserved: amino acid alignment shows high sequence similarity in the nuclear localization signal at the N-terminal region and some similarity in the C-terminal acidic domain. Indeed, TFs containing ERF/AP2DNA-binding domain are widely found in many plants such as Arabidopsis, tomato, tobacco, rice and maize.

Another ABA-independent pathway was identified after the observation that Early Responsive to Dehydration Stress 1 (ERD1) gene transcripts accumulated before any increase of ABA in response to dehydration and high salinity. Promoter analysis of ERD1 revealed TFs belonging to the NAC family and zinc finger homeodomain (ZF-HD) as essential to the activation of the ERD1 gene. The increased drought tolerance may be due both to the reduced transpiration rate (increased stomatal closure) and to an increased ABA sensitivity.

Many genes (e.g. Aquaporin, ERD10, ERD13 and ERF) already described as being involved in plant response to water stress are down-regulated in drought stress. A member of the A. thaliana family of R2R3-MYB TFs, AtMYB61, is also specifically expressed in guard cells in a consistent manner, being involved in the regulation of stomatal aperture.

The strong induction of Stress Responsive –NAC1 (SNAC1) gene expression by drought in guard cells suggests an effect in stomatal closure. It has been reported that modulation of transcription plays an important role in controlling guard cell activity. Recently two MYB-type TFs were identified as regulators of stomatal movements.

## ABA-dependent Pathway

ABA-dependent gene induction during water deficit is controlled by at least five different classes of TFs. The ABA response element (ABRE) with the consensus ACGTGG/TC is bound by basic Leucine Zipper Domain (bZIP-type) TFs. Three Arabidopsis bZIP TFs (AREB1/ABF2, AREB2/ABF4, and ABF3) are expressed in response to water-deficit stress and ABA treatment. Activation of the TFs requires ABA accumulation and the induction of an ABA-responsive protein kinase which activates the TF through phosphorylation.

Other TFs are also involved in ABA regulation of gene expression during cellular water deficit. Three genes encoding a class of TFs that is unique to plants, the NAC domain proteins ANAC019, ANAC055, and ANAC072 are induced by water deficit and ABA treatment. The NAC domain is a 60 bp DNA binding domain that is predicted to form a helix-turn-helix motif.

MYB, MYC and homeodomain TFs, and a family of transcriptional repressors (Cys2/His2-type zinc-finger proteins) are also involved in the ABA response to water deficit. Expression of the drought-inducible gene Responsive to Dehydration 22 (RD22) from Arabidopsis was found to be induced by ABA. The promoter region of RD22 contains MYC (CANNTG) and MYB (C/TAACNA/G) cis-element recognition sites. MYC and MYB TFs only accumulate after an increase of ABA concentration. Over-expression of these TFs result in enhanced sensitivity to ABA and drought tolerance.

## Transcriptional Factor Involved in Response to Flooding Stress

Flooding and submergence are two conditions that cannot be tolerated by most plants for periods of time longer than a few days. These stresses lead to anoxic conditions in the root system. At a critical oxygen pressure, mitochondrial respiration that provides the energy for growth in the photosynthetically inactive roots will decrease, then cease and the cells will die.

Recent reviews on gene expression analysis performed by microarray tools reported as the expression of several transcription factors, such as heat shock factors, ethylene response-binding proteins, MADS-box proteins, AP2 domain, leucine zipper, zinc finger and WRKY factors, increases in response to various regimes of oxygen deprivation in Arabidopsis and rice.

Recently Licausi et al., using a qRT-PCR platform, have identified TFs that are differentially expressed by hypoxic conditions. Among the TFs that have been characterized, members of the AP2/ERF-type family are the most commonly represented in the set of up-regulated TFs, followed by Zinc-finger and basic helix-loop-helix (bHLH-type) TFs, while TFs belonging to the bHLH family are the most commonly represented in the set of down-regulated TFs, together with members from the bZIP and MYB families.

In silico experiments and trans-activation assays shown that some TFs active in flooding stress are able to regulate the expression of hypoxia responsive genes. Particularly, five hypoxia-induced TFs (At4g29190; LBD41, At3g02550;HRE1, At1g72360; At1g69570; At5g66980) from different TF families Zinc Finger, Ligand Binding Domain (LBD) or Lateral Organ Boundary Domain, ERF, DNA binding with one finger (DOF), ARF showed this ability.

Accumulation of ROS is a common consequence of biotic and abiotic stresses, including oxygen deprivation. There is evidence of redox-sensitive TFs, at least one of which might be involved in the adaptive response to low oxygen. ZAT12, a putative zinc finger-containing TF, is recognized as a component in the oxidative stress response signalling network of Arabidopsis, promotes expression of other TFs and the upregulation of cytosolic ascorbate peroxidase 1, a key enzyme in the removal of $H_2O_2$.

Advances have been made in molecular analyses of cDNAs and genes involved in the anaerobic response. Huq and Hodges reported early activation of a rice (Oryza sativa L.) gene by anoxia, the aie (anaerobically inducible early) gene. This gene encodes for a putative protein that shows short stretches of similarities to functionally interesting proteins (e. g. DNA binding proteins and nitric oxide synthase), indicating its putative involvement in signalling.

## Salinity Stress

High salinity is a critical environmental factor that inimically affects large areas of cultivated land. Plant growth, physiological and metabolic processes are affected, resulting in significant reductions in global crop productivity. Exposure to high levels of NaCl not only affects plant water relations but also creates ionic stress in the form of cellular accumulation of $Cl^-$ and, in particular, $Na^+$ ions. Salt stress also changes the homeostasis of other ions such as $Ca^{2+}$, $K^+$, and $NO^{3-}$.

Salt accumulation can modify plant cell plasma membrane lipid and protein composition, cause ion imbalance and hyperosmotic stress and eventually disturb normal growth and development.

In general, high NaCl concentrations affect plant physiology and metabolism at different levels, and at least two main responses can be expected: a rapid protective response together with a long term adaptation response. During initial exposure to salinity, plants experience water stress, which in turn reduces leaf expansion. During long-term exposure to salinity, plants experience ionic stress, which can lead to premature senescence of adult leaves, and thus a reduction in the photosynthetic area available to support continued growth.

Salt tolerance determinants are categorized either as effectors that directly modulate stress etiology or attenuate stress effects, or as regulatory molecules that are involved in stress perception, signal transduction, or modulation of effector function. Genomics studies are focused on gene expression analysis following exposure of plants to high salinity, using salt shock experiments to mimic stresses that affect hydration and ion homeostasis.

The stress-responsive genes can be classified into two classes, i.e. early and delayed response genes. The former are induced quickly and transiently, while the latter are activated more slowly and their expression is sustained. The early response genes encode transcription factors that activate downstream delayed response genes.

When microarray expression profiles of wild type plants, a T-DNA insertion knockout mutant of AtNHX1 (nhx1), and a rescued line (NHX1::nhx1) exposed to both short (12 h and 48 h) and long (one and two weeks) durations of a non-lethal salt stress were investigated, 147 transcripts showed both salt responsiveness and a significant influence of AtNHX1. Fifty-seven of these genes showed differential regulation across all salt treatments, while the rest were regulated as a result of a particular duration.

A large number of genes from a variety of biochemical pathways participate in responses conferring salt tolerance. These pathways include notably those involved in: signal transduction; carbon metabolism and energy production; oxidative stress protection; uptake, exclusion, transport and compartmentalization of sodium ions; modifications of structural components of cell walls and membranes.

Several genes have been identified as functional components in the plant response to salt stress, including those encoding detoxifying enzymes like glutathione peroxidase, $Na^+/H^+$ antiporter AtNHX1, osmolytes such as glycine-betaine and LEA (late embryogenesis abundant protein), flavoprotein AtHAL3, signal mediator $Ca^{2+}$/calmodulin-dependent protein phosphatase and transcription factor Alfin1. Analyses of complete transcriptomes suggest that systems like synthesis of osmolytes and ion transporters and regulation of transcriptional and translational machineries have distinct roles in salt-stress response. In particular, induction of transcripts of specific TFs, RNA-binding proteins, ribosomal genes and translation initiation and elongation factors has been reported to be important during salt stress.

Since not many stress-specific consensus sequences were identified in promoters of stress specific genes to activate or repress transcription, transcription factors must be located in the nucleus, bind DNA and interact with the basal transcription apparatus. Transcription factors involved in stress responses include DRE-related binding factors, leucine zipper DNA-binding proteins, putative zinc finger proteins, myb proteins, bZIP/HD-ZIPs, and AP2/EREBP, interact with promoters of osmotic-regulated genes. Particularly, AP2/ERF domain proteins include the DREB or CBF proteins binding to dehydration response elements (DRE) or C-repeats. A major transcriptional regulatory system is represented by DRE/C-repeat promoter sequences in stress-activated genes and DREBs/CBF factors that control stress gene expression. Several stress-inducible genes such as rd29A, Cor6.6, Cor15a and Kin1 are target genes of DREBs/CBFs in Arabidopsis and contain DRE/C-repeat sequences in their promoters.

Moreover, basic region leucine zipper (bZIP) proteins contain a DNA binding domain rich in basic residues that bind to an ACGT core sequence. One bZIP subfamily has been linked genetically to an ABA response: ABI5 and its homologs, theABREbinding factors (ABFs/AREBs). ABRE binding factors (ABFs)/ABA-responsive element binding (AREBs) proteins respond at the transcriptional and post-transcriptional level to dehydration and salt stress.

Other regulatory intermediates that modulate plant salt stress responses include SOS3 ($Ca^{2+}$-binding protein), SOS2 (Suc nonfermenting-like) kinase, $Ca^{2+}$-dependent protein kinases, and mitogen-activated protein kinases. Genetic and physiological data indicate that SOS3, SOS2, and SOS1 are components of a signal pathway that regulates ion homeostasis and salt tolerance and their functions are $Ca^{2+}$ dependent. In particular, SOS1, encoding a plasma membrane $Na^+/H^+$ antiporter, plays a critical role in sodium extrusion and in controlling long-distance $Na^+$ transport from the root to shoot. This antiporter forms one component in a mechanism based on sensing of the salt stress that involves an increase of cytosolic [$Ca^{2+}$] and reversible phosphorylation with SOS1 acting in concert with SOS2 and SOS3. SOS2 encodes a Suc non-fermenting-like (SNF) kinase, and SOS3 encodes a $Ca^{2+}$-binding protein with sequence similarity to the regulatory subunit of calcineurin and neuronal $Ca^{2+}$ sensors. In yeast, co-expression of SOS1, SOS2, and SOS3 increases the salt tolerance of transformed yeast cells much more than expression of one or two SOS proteins, suggesting that the full activity of SOS1 depends on the SOS2/SOS3 complex.

Several studies have shown that reactive oxygen species (ROS) and oxidative stress may be mediating at least some of the toxic effects of NaCl on legumes and other vascular plants. ROS are predominantly generated in the chloroplast by direct transfer of excitation energy from chlorophyll to produce singlet oxygen, or by univalent oxygen reduction at photosystem I, in the Mehler reaction and to some extent in mitochondria. ROS have the potential to interact non-specifically with many cellular components, triggering peroxidative reactions and causing significant damage to proteins, lipids, and nucleic acids. To cope with ROS and to maintain redox homeostasis, living organisms evolved antioxidant defense systems, comprised of enzymatic and non-enzymatic components, which normally maintain ROS balance within the cell. Major nonenzymatic antioxidants include ascorbate (vitamin C) and glutathione in plants, although tocopherol (vitamin E), flavonoids, alkaloids, and carotenoids can also act as antioxidants.

Intracellular ROS can also influence the ROS induced MAPK signal pathway through inhibition of phosphatases or downstream transcription factors.

## Chilling and Cold Stress: Gene Regulation and Transcriptional Factor

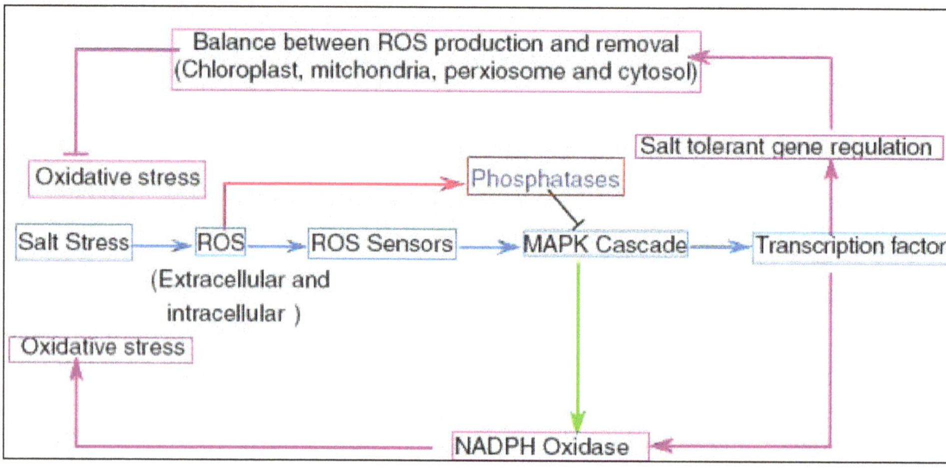


ROS signal transduction pathway under salt stress.

Cold stress prevents the expression of full genetic potential of plants owing to its direct inhibition of metabolic reactions and, indirectly, through cold-induced osmotic (chilling-induced inhibition of water uptake and freezing-induced cellular dehydration), oxidative and other stresses. Cold stress, which includes chilling (<20°C) and freezing (<0°C) temperatures, adversely affects the growth and development of plants. Chilling and freezing are stresses that show different effects on plants: the first leads to slow biochemical reactions, such as enzyme and membrane transport activities; the second leads to ice crystal formation that can cause the disruption of cell membrane system.

A large number of studies have used a transcriptional profiling approach to identify genes in Arabidopsis that respond to cold (4°C) and chilling (13°C) temperatures. Results have shown that plants respond to low temperatures by altering mRNA levels of a large number of genes belonging to different and independent pathways. The quantitative and qualitative difference in transcriptional response to low temperature suggests the presence in higher plants of different molecular mechanisms to cold-stress response.

The cold induction of genes involved in calcium signalling, lipid signalling or encoding receptor-like protein kinases are also affected by the ice1 mutation.

Controlled proteolysis of transcriptional regulators also plays an important role in shaping the cold-responsive transcriptome in plants.

TFs that bind to the DRE/CRT are named DREB1/CTR-binding factor (CBF) and DREB2. Cold stress induces the expression of AP_2/ERF family TFs, that is, CBFs, which can bind to cis-elements in the promoters of COR genes and activate their expression. CBFs regulate the expression of genes involved in phosphoinositide metabolism, transcription, osmolyte biosynthesis, ROS detoxification, membrane transport, hormone metabolism and signalling and many others with known or presumed cellular protective functions.

The first isolated cDNAs encoding DRE binding proteins were DREB1A and DREB2A from Arabidopsis and then, DREB genes have been isolated from a wide variety of plants. In wheat and barley, a number of CBF homologs have been mapped to low temperature QTLs, Fr-2 chromosomal region. Thus, it is clear that the DREB1/CBF regulon is ubiquitous within higher plants.

Expression of DREB1 genes was extensively investigated in various crops with regard to different abiotic stresses. It was found that the expression of AtDREB1 gene is induced by cold, but not by dehydration, or high salt stress. Similarly, CBF genes also showed high expression in response to low temperature treatment and its transcript was detectable after 30 min of exposure to 4 C, and showed maximum expression at 1 h. Indeed, CBF regulon could be sub-regulated by cold-responsive transcription factor genes RAP2.1 and RAP2.7 as shown by microarray analysis of transgenic Arabidopsis plants ectopically expressing CBFs.

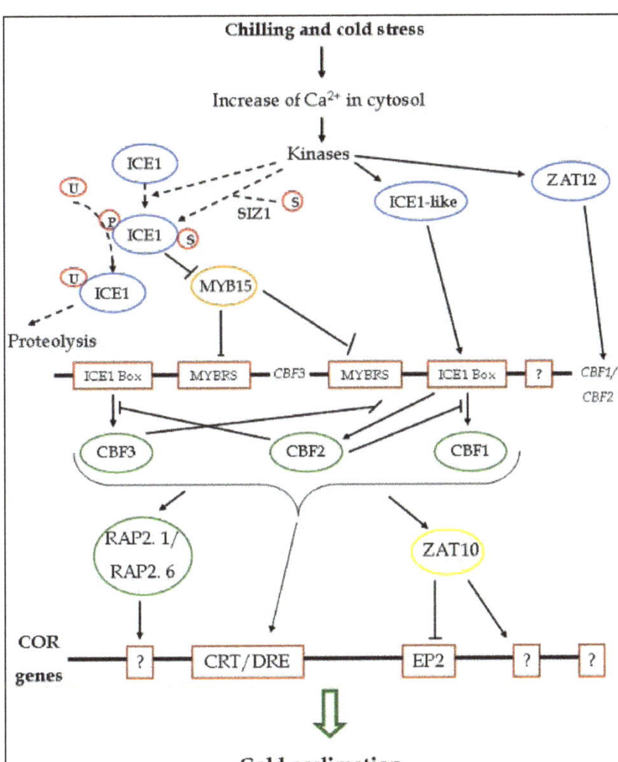

Cold-responsive transcriptional network in Arabidopsis. CBFs regulate the expression of COR genes that confer cold tolerance. CBFs might cross-regulate the each other's transcription. CBFs induce the expression of ZAT10 which might downregulate the expression of COR genes. Costitutive expressed ICE1 is actived throught sumoylation and phosphorylation induced by cold stress. ICE1 actived induce the transcription of CBFs and reprime MYB15. The expression of CBFs is negatively regulated by MYB15 and ZAT12. HOS1 mediates the ubiquitination and proteolysis of ICE1, thus negatively regulates CBF regulons. Lines ending with bar indicate negative regulation; question mark (?) indicate unknown cis-elements; broken arrows indicate post-translational regulation; solid arrows indicate activation; lines ending with bar indicate negative regulation.

In Arabidopsis, ICE1 (Inducer of CBF Expression1), a MYC-type bHLH TF, can bind to MYC recognition elements in the CBF3 promoter and is important for the expression of CBF3 during cold acclimation. ICE1 is constitutively expressed and localized in the nucleus, but it induces expression of CBFs only under cold stress. This suggests that cold stress-induced post-translational modification is necessary for ICE1 to activate downstream genes in plants.

Two important post-translational protein modifications are the ubiquitination and the sumoylation. Ubiquitination is mediated by High Expression of Osmotically Responsive1 (HOS1). For HOS1 encodes for a RING finger ubiquitin E3 ligase that physically interacts with ICE1 and mediates the ubiquitination of ICE1 to regulate negatively the expression of ICE1 target genes and is thus critical for the de-sensitization of plant cells to cold stress. Sumoylation is induced by SUMO (Small Ubiquitin-related Modifier) proteins that are conjugated to proteins substrates in a process dependent on SUMO E3 ligases. Sumoylation might protect target proteins from proteasomal degradation preventing the ubiqiutination.

## Heavy Metal Accumulation and Metal Stress

Uptake of excess metal ions is toxic to most plants. Phytotoxicity of heavy metals can be attributed to symplastic accumulation of heavy metals, particularly in the plasmatic compartments of the cells, such as the cytosol and chloroplast stroma. Metal-induced changes in development are the result of either a direct and immediate impairment of metabolism or signalling processes that initiate adaptive or toxicity responses that need to be considered as active processes of the organism. The detoxification of heavy metals by plants is achieved by uptake and translocation, sequestration into the vacuole and metabolization, including oxidation, reduction or hydrolysis and conjugation with glucose, glytanyl cysteine syntase (GSH) or amino acids.

So, in order to determine genes involved in response to heavy metal, recently, several studies, based on use of A. thaliana as model plant, performed the analysis of global gene expression after exposure to salts of lead (Pb) and cadmium (Cd). The analysis revealed 65 and 338 up and down-regulated genes by Cd and 19 and 76 by Pb. Particularly, it was found that ABC transporters were differentially regulated after Cd treatments, suggesting for some plant ABC transponders a key role in glutathione-Cd or phytochelatins-Cd complex transport both into cellular compartments and outside of the cell.

Subsequently studies performed on Arabidopsis, using microarray tools, demonstrated that exist a complex regulatory network which differentially modulates gene expression in a tissute-specific manner. Responses observed in roots included the induction of genes involved in sulphur assimilation-reduction and glutathione metabolism. Therefore, it was suggested that plants activate the sulphur assimilation pathway by increasing transcription of related genes to provide an enhanced supply of glutathione for phytochelatin biosynthesis.

Non specific defense mechanisms include accumulation of osmolytes, antioxidants, aminoacids and changes in hormonal balances.

The significance of glutathione and the metal-induced phytochelatins (PCs) in heavy metal tolerance has been summarized intensely in excellent reviews. Depletion of glutathione appears to be a major mechanism in short-term heavy metal toxicity and in accordance with this hypothesis, a good correlation between glutathione contents and tolerance index was observed with 10 pea genotypes differing in Cd sensitivity.

In roots, after Cd exposure, three categories of genes were identified from transcriptome analysis: (1) common responses conserved across species; (2) metallophyte-specific responses representing candidate genes for Cd hypertolerance; (3) specific responses to Cd.

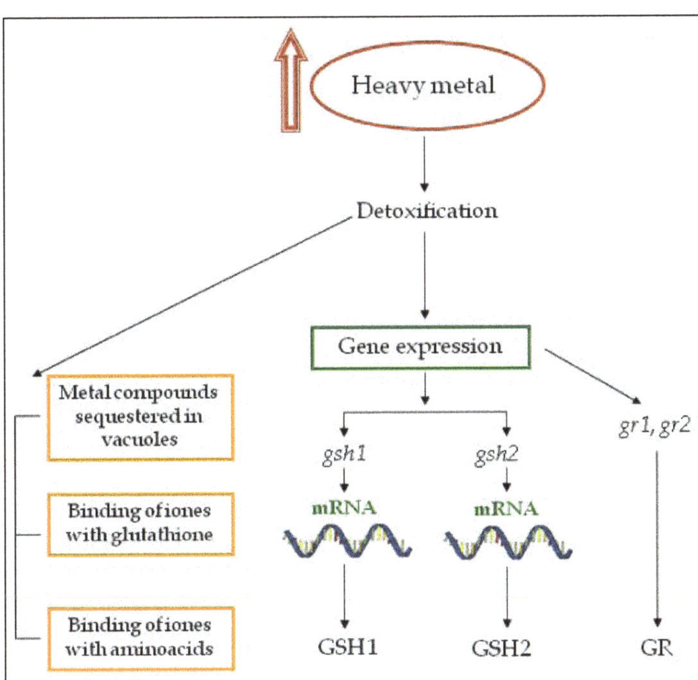

Response of plant cell to toxic levels of heavy metals. The synthesis of phytochelatins (PCs) accompanies with decrease in cell glutathione pool and increase in the activities of glytanyl cysteine syntase (GSH1), glutathione synthetase (GSH2) and glutathione reductase (GR). The elevated activities of GSH1, GSH2 and GR is correlated with enhanced expression of corresponding genes gsh1, gsh2, gr1 and gr2.

In leaves, instead, was reported an early induction of several genes encoding enzymes involved in the biosynthesis of phenylpropanoids.

## High Light Stress

Light plays a critical role in regulating plant growth and development through the modulation of expression levels of light-responsive genes that regulate developmental and metabolic processes. Light signals are perceived through at least four distinct families of photoreceptors, which include phytochromes (Phy), cryptochromes, phototropins and unidentified ultraviolet B (UVB) photoreceptor(s). For each developmental response, more than one photoreceptor can contribute to the perception of light signals, indicating that signal integration points for different light signals must exist in transcriptional hierarchies. Light can modulate photoreceptor activity by inducing changes that alter their cellular localization. The best characterized light receptor is Phy, which exists in two photochemically interconvertible forms, Pr and Pfr, and is encoded by a small family of genes in angiosperms. Phytochromes are synthesized in the inactive Pr form, that absorbs red light, (660 nm), and are activated on light absorption by conversion to the biologically active Pfr form, that absorbs far-red light (730 nm). The photoconversion of phytochromes results in their translocation from the cytoplasm into the nucleus, which is crucial for allowing them to interact with transducers in initiating downstream transcriptional cascades.

The responses of plants to light are complex: seed germination, seedlings photomorphogenesis, chloroplast development and orientation, photodinesis, stem growth, pigment biosynthesis, flowering and senescence. Collectively these processes are known as photomorphogenesis.

Besides excess light, a range of abiotic environmental conditions such as $O_3$, salt, toxic metals, and temperature can induce increased production of ROS by limiting the ability of a plant to utilize light energy through photosynthesis. Exposure of a plant to light exceeding what is utilized in photochemistry leads to inactivation of photosynthetic functions and the production of reactive oxygen species (ROS) such as hydrogen peroxide ($H_2O_2$), superoxide ($O_2^-$), hydroxyl radicals, and singlet oxygen. Indeed, high light drove change in the redox potential of plastoquinone (PQ) regulating the expression of two cytosolic peroxidases during HL stress. Furthermore, the redox state of PQ has been shown to be involved in the expression of chloroplast encoded genes.

Classical genetic and molecular approaches have identified various regulators downstream of photoreceptors. Many of these encode TFs, as well as kinases, phosphatases and degradation-pathway proteins. Although some of these regulators are specific for light quality, others regulate signal transduction networks in response to various light signals, representing potential signal integration points. Several basic post-translational mechanisms are involved in regulating TF activities and the subcellular localization in response to light. The phosphorylation of TFs is a common modification that can influence their ability to bind to promoters. For example, the level of G-Box Binding Factor 1 (GBF1) is constant, but its affinity for the G-box is modulated by its phosphorylation status: its phosphorylation by nuclear Casein Kinase II (CKII) enables G-box binding.

In the dark, some TFs that positively regulate gene expression in response to light, such as Long After Farred Light 1 (LAF1), are ubiquitylated by Constitutive Photomorphogenic 1 (COP1), a ring-finger-type ubiquitin E3 ligase. In darkness, COP1 acts as E3 ligase in the nucleus, targeting TFs like Long Hypocotyl5 (HY5) and LAF1 to degradation via the 26S proteasome. Upon exposure to light, COP1 migrates from the nucleus to the citosol. HY5, a bZIP transcription factor that is one of the key regulators of cryptochrome and phytochrome controlled photomorphogenesis, is an important component of the UVB-induced signalling network. UVB promotes rapid transcriptional activation of HY5 (and its interacting partner Long Hypocotyl5-Like [HYH]) independently of all known photoreceptors, and loss of HY5 results in the impairment of the transcriptional induction of a subset of UVB-responsive genes. Taken together, these observations demonstrate that UVB up-regulates HY5 transcription by yet-unknown signalling pathways, and that the signalling cascades that mediate responses to visible light and long-wavelength UVB (300–320 nm) use shared components. Additional studies suggested that HY5 also regulates the transcription of several photosynthesis-related genes, such as the ribulose bisphosphate carboxylase small subunit (RbcS1A). Given that HY5 appears to regulate the expression of several Arabidopsis genes known to respond to abiotic stress conditions (e.g. CBF1, DREB2A, RD20 and MYB59), it is inferred that HY5 could also be involved in the regulation of photosynthesis by adverse environmental conditions.

In vitro analysis showed that HY5 directly binds to the promoters of several light-inducible genes and a recent chromatin immuno-precipitation analysis in combination with a whole-genome tiling microarray revealed that HY5 binds directly to a large number of genomic sites, mainly at the promoter regions of annotated genes. HY5 interacts specifically with the G-box (CACGTG) and is required for normal control by light of promoters bearing this sequence.

Recently, some review showed as DNA cis-elements responsible for light regulated transcription are located within 5' upstream sequences.

The evolution of regulatory sequences, which determine where, when, and the level at which genes are transcribed, has been largely neglected. In the case of the photosynthesis-associated nuclear genes (PhANGs) from higher plants, interesting evolutionary aspects of the molecular mechanisms by which transcription is activated by light receptors (e.g. phytochrome) could be addressed through the comparative analysis of promoter sequences. For instance, why does light profoundly affect transcription of

PhANGs in monocotyledonous and dicotyledonous plants, while PhANG promoters in conifers, ferns, and mosses are either light insensitive or, at most, weakly photoresponsive.

Light-responsive Transcriptor Factors (TFs) have been identified through screens for light-responsive cis-element (LRE)-binding proteins and through genetic analyses of mutants that are deficient in their response to specific types of light. A combination of various methods has been used to identify these LREs. Such analyses have been successfully perdormed in identifying cis-acting elements involved in the light responsiveness of PhANGs, such as the G-box and I-box elements from rbcS genes and the GATA motifs of Lhcb1 genes.

Although many LREs and their binding proteins have been identified, no single element is found in all light-regulated promoters, suggesting a complex light-regulation network and a lack of a universal switch. Sequence heterogeneity of regulatory elements may be functionally overcome if multiprotein regulatory complexes facilitate binding to imperfect target sites. The individual elements found within a multipartite cis-regulatory region are termed phylogenetic footprints (PFs); they share high conservation over a segment of 6 contiguous base pairs in alignments of orthologous upstream sequences and represent potential binding sites for transcription factors.

The "phylogenetic-structural method" is based on the search of "homologous" (rather than "similar") DNA sequences of a functionally characterized promoter. Two sequences are homologous when they share common ancestry, regardless of the degree of similarity between them.

# ABIOTIC STRESS SIGNALING AND RESPONSES IN PLANTS

Plants live in constantly changing environments that are often unfavorable or stressful for growth and development. These adverse environmental conditions include biotic stress, such as pathogen infection and herbivore attack, and abiotic stress, such as drought, heat, cold, nutrient deficiency, and excess of salt or toxic metals like aluminum, arsenate, and cadmium in the soil. Drought, salt, and temperature stresses are major environmental factors that affect the geographical distribution of plants in nature, limit plant productivity in agriculture, and threaten food security. The adverse effects of these abiotic stresses are exacerbated by climate change, which has been predicted to result in an increased frequency of extreme weather. How plants sense stress signals and adapt to adverse environments are fundamental biological questions. Further, improving plant stress resistance is critical for agricultural productivity and also for environmental sustainability because crops with poor stress resistance consume too much water and fertilizers and thus greatly burden the environment.

For drought and salt stress, it is helpful to distinguish the primary stress signals from secondary signals caused by too little water or too much salt. The primary signal caused by drought is hyperosmotic stress, which is often referred to simply as osmotic stress because a hypo-osmotic condition typically is not a significant problem for plant cells. Salt stress has both osmotic and ionic or ion-toxicity effects on cells. The secondary effects of drought and salt stresses are complex and include oxidative stress; damage to cellular components such as membrane lipids, proteins, and nucleic acids; and metabolic dysfunction. Although some cellular responses ensue from primary stress signals, others arise primarily from secondary signals. Thus drought and salt have unique and overlapping signals. An important feature of drought and salt stress is that the hyperosmotic signal causes the accumulation of the phytohormone abscisic acid (ABA), which in turn elicits many adaptive responses in plants.

## Stress Sensing and Putative Sensors

Because plants show specific changes in gene expression, metabolism, and physiology in response to different environmental stress conditions, it is safe to presume that plant cells must be capable of sensing various environmental signals. How-ever, despite much effort, only a few putative sensors have been identified. The difficulty seems to lie in the functional redundancy in genes encoding sensor proteins such that dysfunction in one gene does not cause significant phenotypes in stress responses. Alternatively, a sensor may be essential for the plant, such that loss-of-function mutants are lethal, precluding further studies. In addition, demonstrating that a protein or other macromolecule functions as a sensor for a physical signal (e.g., change in osmotic pressure, ion concentration, or temperature) is technically very challenging, and direct experimental evidence is absent even for widely accepted osmosensors or temperature sensors in bacterial, yeast, or mammalian systems.

A putative sensor for hyperosmotic stress is the Arabidopsis OSCA1 (reduced hyperosmolality-induced calcium increase 1. Osmotic stress agents, high salt, cold, and heat as well as oxidative stress, heavy metals, and ABA, can cause increases in the cytosolic free calcium concentration in plants, which can be detected by researchers using genetically encoded aequorin or other calcium reporters. Arabidopsis osca1 loss-of-function mutants display a reduced calcium spike compared to wild-type plants when treated with osmotic stress agents such as sorbitol and mannitol. OSCA1 encodes a plasma-membrane protein that forms hyperosmolality-gated calcium-permeable channels. Because these mutant plants do not display any drought or salt stress phenotypes, the physiological significance of OSCA1 under the stress conditions is unclear. We do not know how OSCA1 senses osmotic stress, which is expected to reduce turgor and thus may affect membrane stretching and membrane–cell wall interactions. A number of mechano-sensitive channels, including TRP, MscS-like, Piezo, DEG/ENaC, and K2P, are known in non-plant systems. Animal TRP channels are well-known calcium channels that can sense changes in the membrane caused by fluctuations in temperature or osmotic pressure. Plants lack TRP and DEG/ENaC genes but contain a family of MscS-like proteins and one Piezo homolog. One of the MscS-like proteins in Arabidopsis, MSL8, is required for pollen to survive hypo-osmotic shock during hydration, suggesting MSL8 as a sensor of hypo-osmotic stress-induced membrane tension. Plants also have a large family of cyclic nucleotide-gated channels (CNGCs) as well as a family of glutamate receptor-like (GLR) channels that are potentially very important in generating cytosolic $Ca^{2+}$ signals under stress.

Another putative stress sensor is COLD1, which was recently reported to mediate cold stress sensing in rice. COLD1 is required for chilling (0–15°C) tolerance in the Nipponbare sub-species of rice. COLD1 is a transmembrane protein in the plasma membrane and ER and interacts with RGA1, the alpha subunit of the sole heterotrimeric G protein in plants. The authors speculated that COLD1 may regulate a calcium channel or is itself a cold-sensing calcium channel. It is unclear how COLD1-mediated calcium signaling leads to chilling tolerance.

Both cold and heat can change the fluidity of cellular phospholipid membranes. This change may be sensed by integral membrane proteins, including channels and various other transporters and membrane-anchored receptor like kinases (RLKs). Heat stress may also be sensed by molecular chaperones that bind to misfolded proteins caused by heat denaturation. The binding of misfolded proteins would release heat-stress transcription factors from the chaperones, allowing the heat-stress transcription factors to activate heat-responsive genes. Recently, H2A. Z-containing nucleosomes were suggested to be a thermosensory apparatus in plants and yeast. The authors proposed that H2A. Z-containing nucleosomes wrap DNA more tightly than H2A-containing nucleosomes and that elevated temperatures increase the expression of heat shock proteins (HSPs) and other genes by dissociating H2A. Z nucleosomes to make DNA more accessible to Pol II transcription. This is an attractive hypothesis that remains to be further tested.

## Sensing of Organellar Stress

Stress sensing is often compared to ligand perception, so it is frequently thought to occur on the cell surface or at the cell membrane. Then, the signal would be relayed to various subcellular locations such as the nucleus. Theoretically, physical stress sig nals, particularly temperature signals, may be sensed anywhere in the cell, as long as the stress signal causes a change in the status of the cellular component (protein, DNA, RNA, carbohydrate, or lipid) or compartment (e.g., metabolic reaction in a compartment), and as long as that change then affects other cellular components or activities. Stress-induced changes in protein folding in the ER, referred to as ER stress, are now widely recognized as an important cellular response to stress conditions. Similarly, there is stress associated with other organelles such as the chloroplast, mitochondrion, peroxisome, nucleus, and cell wall, and the stress-generated signals from all organelles are integrated to regulate stress-responsive gene expression and other cellular activities to restore cellular homeostasis.

## ER Stress

Both biotic and abiotic stress can cause protein misfolding or the accumulation of unfolded proteins, which is sensed as ER stress by specific sensor proteins in the ER membrane. This sensing leads to the expression of genes encoding chaperones and other proteins important for enhancing protein folding capacity, ER-associated degradation (ERAD), or protein translation suppression to reduce the amount of synthesized proteins loaded to ER via the PKR-like ER eIF2a kinase. These changes help restore ER homeostasis, i.e., the equilibrium between protein-folding demands and folding capacity, and are known as the unfolded protein response (UPR), a conserved stress response in eukaryotes. Two main types of sensors of ER stress have been identified in plants: ER membrane-associated transcription factors and an RNA-splicing factor. The basic leucine zipper bZIP28 may sense heat and other ER stress agents through its interaction with the chaperone protein BIP (binding immunoglobulin protein) in the ER. Unfolded or misfolded proteins accumulate under stress, and these proteins can interact with BIP, which releases bZIP28 for transport to the Golgi, where it is cleaved. Its cytosolic portion then relocates to the nucleus to activate the expression of stress-response genes to restore ER homeostasis. bZIP28 may also sense other changes that promote its release from BIP, including alterations in energy charge levels and redox status or interactions between BIP and the DNAJ domain-containing protein chaperone. bZIP17 can be activated by salt stress in a similar manner. In addition, several ER or plasma-membrane-associated NAC transcription factors can be activated by ER stress and contribute to UPR. The second type of ER stress sensor in plants is IRE1, a splicing factor conserved from yeast to metazoans. Presumably, plant IRE1 proteins bind to unfolded proteins and sense ER stress in a manner similar to that of their yeast homolog. Activated IRE1 in Arabidopsis recognizes and splices bZIP60 mRNA and perhaps other target mRNAs. The splicing of bZIP60 mRNA by IRE1 results in a bZIP60 variant that can enter the nucleus to activate UPR genes.

## Chloroplast Stress

The chloroplast is an organelle where photosynthetic electron transport and many metabolic reactions take place; the metabolic balance in chloroplasts is easily perturbed by environmental stresses. A disturbance in chloroplast homeostasis is communicated to the nucleus through retrograde signals so that all cellular activities can be adjusted and coordinated according to the ability of the stressed chloroplast to supply sugar and other compounds.

Stress Sensing and Signaling in Different Cell Organelles (i) Model of dispersed stress sensing by organelles. Stress causes perturbations in various organelles, generating signals that are integrated to regulate nuclear gene expression and other cellular activities, which helps to restore cellular homeostasis. (ii) ER stress sensing and signaling. (iii) Chloroplast stress sensing and signaling. Dashed lines indicate postulated regulation.

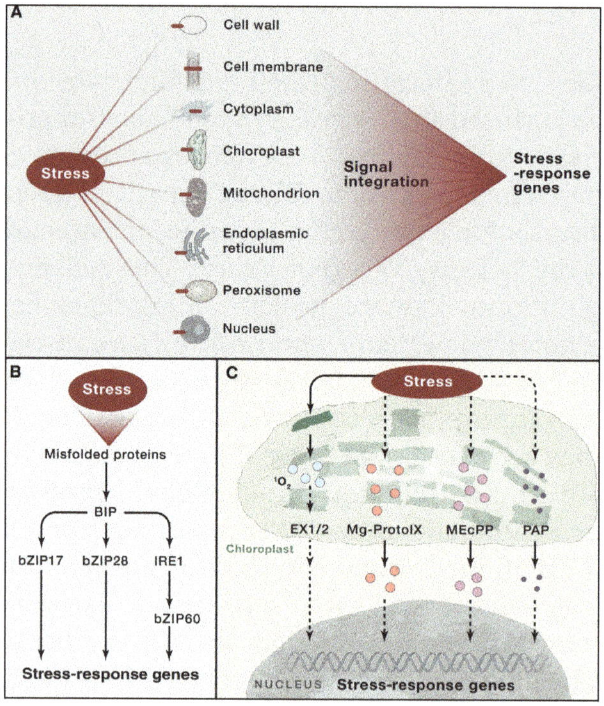


The chloroplast is a major site for the production of ROS including superoxide anion, hydrogen peroxide, hydroxyl radical, and singlet oxygen. Various environmental stresses, particularly high light stress, exacerbate ROS production, which overwhelms ROS-managing systems and generates various secondary messengers.

Singlet oxygen triggers a signaling pathway that requires EXECUTER (EX1) and EX2, two nuclear-encoded proteins in the thylakoid membrane of chloroplasts. The Arabidopsis fluorescent (flu) mutant produces a burst of singlet oxygen at dark-to-light transitions because it accumulates the chlorophyll precursor protochlorophyllide. The accumulation of singlet oxygen triggers dramatic changes in nuclear gene expression and causes chlorosis and cell death in wild-type but not in ex mutants. Whether EX1 and EX2 directly sense singlet oxygen and how these proteins relay the singlet oxygen signal to the nucleus need resolution. Singlet oxygen-triggered signal also occurs independently of EX1 and EX2, possibly through non-enzymatic oxidative breakdown products of b-carotene.

High light and other stresses can also cause increases in the level of the plastid metabolite methylerythritol cyclo-diphosphate (MEcPP), a precursor of isoprenoids. MEcPP functions as a retrograde signal to activate the expression of stress-responsive nuclear genes that encode plastid proteins. Another plastid metabolite important for stress responses is the phosphonucleotide 30-phosphoadeno-sine 50-phosphate (PAP). PAP levels increase following drought and high light stress. SAL1/FRY1 is a bifunctional phosphatase that can dephosphorylate both inositol phosphates and PAP (to AMP), and its dysfunction causes PAP accumulation. PAP inhibits 50 to 30 exoribonucleases, contributing to the enhanced expression of a subset of drought and high light-responsive genes and to altered drought resistance. In addition to PAP and MEcPP, other chloroplast metabolites, such as the tetrapyrroles and oxidative breakdown products of b-carotene, have been proposed as retrograde signals. How stress modulates the levels of MEcPP, PAP, tetrapyrroles, and other retrograde signaling metabolites is not well understood.

## Mitochondrion and Peroxisome Stress

Like chloroplasts, mitochondria and perhaps even peroxisomes can generate retrograde signals that are important for stress re-sponses. Mitochondria and peroxisomes produce ROS and many metabolites, some of which may serve as retrograde signals. Dysfunction in a mitochondrial DEXH box RNA helicase or mitochondrial pentatricopeptide repeat protein causes ROS accumulation and altered responses to the plant stress hormone ABA. A defect in the mitochondrial complex I also enhances ROS accumulation and causes the mutant plants to have reduced expression of cold-responsive genes and to exhibit chilling and freezing sensitivity. Similarly, mutations in CHY1, which encodes a peroxisomal beta-hydroxyisobutyryl (HIBYL)-CoA hydrolase needed for valine catabolism and fatty acid beta-oxidation, also cause ROS accumulation and impair cold-responsive gene expression and freezing tolerance. ROS signals from mitochondria and peroxisomes may modulate cold stress responses by affecting calcium signaling.

## Cell-wall Stress

In plants, the primary cell wall consists of cellulose fibrils interconnected by hemicellulose tethers, such as xyloglucan and ara-binoxylan, and embedded in a pectin gel. The wall also contains phenolics; peroxidases, pectin esterases, and other enzymes; extensins, expansins, and other proteins; and $Ca^{2+}$. Salt, drought, and other osmotic stress treatments can cause ROS accumulation and other changes in the cell wall. Accumulation of ROS can cause crosslinking of phenolics and cell-wall glycoproteins such as extensins, resulting in cell-wall stiffening. On the other hand, stress upregulates the expression of expansins and xyloglucan-modifying enzymes that can remodel the wall. In addition, salt stress can also cause the wall to lose $Ca^{2+}$. Recently, cell-wall stress has been proposed to trigger dedicated signaling pathways analogous to the fungal cell-wall integrity (CWI) pathway. In yeast, cell-wall stress is sensed by the wall-associated plasma-membrane proteins Wsc1-3, Mid2, and Mtl1. How these sensors detect cell-wall deformations is not known. Because these proteins all contain a large O-mannosylated extracellular domain rich in serine and threonine residues, these proteins may function as mechanosensors. Downstream of these sensors are GDP/GTP exchange factors, small GTPase Rho1, protein kinase C, and a MAPK cascade. The identities of plant cell-wall stress sensors are unclear. However, plants have hundreds of RLKs and other kinases with an extracellular domain, a transmembrane domain, and a cytoplasmic kinase domain, so many candidates exist to sense cell-wall perturbations.

Cell-wall perturbations greatly affect plant stress resistance. A subtle cell-wall defect caused by dysfunction in the putative pectin biosynthesis enzyme AtCSLD5 in the Arabidopsis sos6 (salt overly sensitive 6) mutant results in oxidative stress and greatly increased sensitivity to osmotic treatments and salt and drought stress. Recently, Endler et al. identified two proteins in the cellulose synthase complex that help connect the complex to microtubules and are important for plant growth under salt stress. Much remains to be elucidated about stress-induced physical and chemical alterations in the cell wall, how these alterations are sensed and signals are transduced, and the output of the CWI pathway in plants.

## Ionic Stress Signaling

Soil salinity affects a substantial percentage of cultivated land and is a significant factor limiting agricultural productivity worldwide. High salt levels cause ion toxicity (mainly $Na^+$), hyperosmotic stress, and secondary stresses such as oxidative damage. It is not known how $Na^+$ is sensed in any cellular system. In the yeast Saccharomyces cerevisiae, the calcineurin pathway plays a major role in $Na^+$ stress signaling and tolerance. $Na^+$ stress-triggered cytosolic calcium binds to the EF-hand calcium-binding proteins calmodulin and the B subunit of calcineurin (CnB). $Ca^{2+}$-CnB and $Ca^{2+}$-calmodulin activate the phosphatase catalytic subunit of calcineurin, CnA. The activated phosphatase dephosphorylates the zinc finger transcription factor CRZ1, which then moves to the nucleus to activate the expression of ENA1 and other target genes. ENA1 encodes a $Na^+$-ATPase that pumps the toxic $Na^+$ out of the cell, thus restoring ion homeostasis. Plant genomes do not encode any calcineurin proteins, even though the name calcineurin B-like (CBL) has been widely used to refer to a family of plant EF-hand calcium-binding proteins. Plants instead use a calcium-dependent protein kinase pathway known as the SOS pathway for salt stress signaling and $Na^+$ tolerance. In this pathway, the EF-hand calcium-binding protein SOS3 senses the cytosolic calcium signal elicited by salt stress. SOS3 interacts with and activates SOS2, a serine/threonine protein kinase. SOS3 is preferentially expressed in the root, and an SOS3 paralog, SCaBP8/CBL10 mainly expressed in the shoot, performs a role equivalent to that of SOS3. The activated SOS2 phosphorylates and activates SOS1, a $Na^+/H^+$ antiporter at the plasma membrane.

SOS1 is expressed in root epidermal cells and xylem parenchyma cells, so that activated SOS1 can extrude $Na^+$ into the soil solution and load $Na^+$ into the xylem for long-distance transport to leaves by the transpirational stream. The long-distance role of SOS1 is only partly understood because SOS1 is also expressed in the xylem parenchyma in leaves where its function is unclear. Perhaps there is a Casparin strip-like structure in the xylem parenchyma cells in leaves that prevents the xylem stream from directly entering the apoplastic space of mesophyll cells. As a consequence, SOS1 may function to extrude $Na^+$ from the xylem parenchyma cells into the apoplastic space of mesophyll cells. A prediction of this model is that SOS1 may preferentially localize on the side of the cell facing the mesophyll apoplastic space. HKT1 is another important transporter with a critical role in long-distance $Na^+$ transport. HKT1 in Arabidopsis is a $Na^+$ importer also expressed in xylem parenchyma cells and other cells in the vasculature throughout the plant. In roots, HKT1 may unload $Na^+$ from the xylem to restrict the amount of $Na^+$ in the transpirational stream. In leaves, HKT1 is proposed to load $Na^+$ into the phloem for recirculation back to the root. Dysfunction in any of the SOS genes, including SOS1, greatly increases the sensitivity of the mutant plants to salt stress. In fact, the SOS genes were discovered because of the SOS (salt-overly-sensitive) phenotype of these mutants. Mutations in HKT1 cause obvious salt-stress sensitivity in transpiring plants but can suppress the salt sensitivity of sos mutants grown in culture media where there is minimal transpiration. How the apparently antagonistic functions of SOS1 and HKT1 are coordinated remains an important question in salt-stress research. It is unclear whether SOS1 and HKT1 are expressed in exactly the same cells in the vascular system. Overall, SOS1 seems to promote $Na^+$ transport from roots to leaves, whereas HKT1 seems to restrict this movement. The importance of these two transporters in long-distance $Na^+$ transport may depend on the severity of salt stress. Under mild salt stress, it may be advantageous for the plant to activate SOS1 in xylem parenchyma cells to deliver more $Na^+$ to leaves where $Na^+$ can be stored in the large vacuoles of mesophyll cells for cellular osmotic adjustment to facilitate growth.

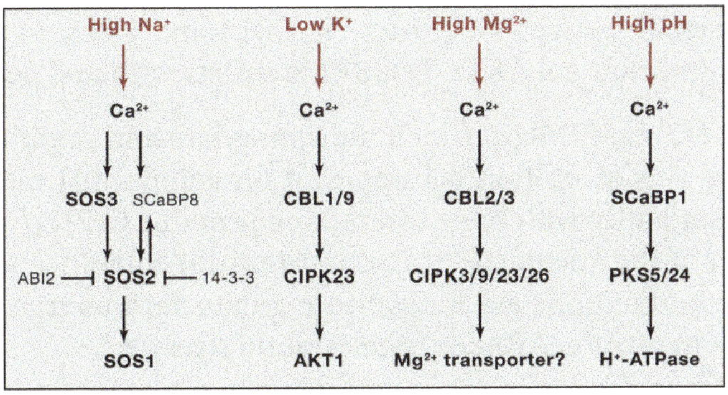


The $Ca^{2+}$-CBL-CIPK Module Mediates Signaling of Various Ionic Stresses High $Na^+$, low $K^+$, excess $Mg^{2+}$, and high pH (low $H^+$) conditions cause cytosolic $Ca^{2+}$ signals, which activate SOS3 (CBL4)/SCaBP8 (CBL10)-SOS2 (CIPK24), CBL1/9-CIPK23, CBL2/3-CIPK3/9/23/ 26, and SCaBP1 (CBL2)-PKS5/24 (CIPK11/14) to phosphorylate and regulate the activity of SOS1 ($Na^+/H^+$ antiporter), AKT1 ($K^+$ channel), a putative $Mg^{2+}$ transporter, and $H^+$-ATPase, respectively. Also shown are ABI2 and 14-3-3 inhibition of SOS2 and SOS2 regulation of SCaBP8 by phosphorylation. Arrows indicate activation, and bars indicate inhibition.

Under severe salt stress, however, high levels of $Na^+$ delivered to leaves may exceed storage capacities, and thus $Na^+$ must be unloaded from the root xylem and also recirculated back from leaves to roots using HKT1. Several SOS3-like calcium-binding proteins are phosphorylated by their interacting SOS2-like protein kinases, and the phosphorylation appears important for the activation of the kinases by the calcium-binding proteins. In the resting state or under non-stress conditions, SOS2 interacts with 14-3-3 proteins to ensure that SOS2 is inactive. In addition, SOS2 interacts with the type 2C protein phosphatase ABI2, which may keep SOS2 inactive.

The SOS pathway was the first abiotic stress-signaling pathway established in plants. The central signaling component, SOS2, represents a large family of protein kinases whose catalytic domain is similar to the yeast sucrose non-fermenting 1 (SNF1) and mammalian AMP-activated kinase (AMPK). In Arabidopsis, these proteins are generally referred to as SNF1-related kinases (SnRKs). SnRKs have 3 members in subfamily 1 (SnRK1s), 10 members in subfamily 2 (SnRK2s), and 25 members in subfamily 3 (SnRK3s). Each of the 25 SnRK3s (aka PKSs or CIPKs), which include the founding member SOS2, interacts with one or more members of the family of 10 SOS3-like calcium-binding proteins (SCaBPs, aka CBLs). The interaction is mediated through a common motif known as FISL in the N-terminal regulatory region of the kinases. Deletion of the FISL motif or the entire regulatory region leads to constitutive activation of the kinases. The large number of possible SCaBP/CBL-PKS/CIPK combinations suggests that the $Ca^{2+}$-SOS3-SOS2 signaling principle is widely used in plants. Indeed, the CBL-CIPK module is important for various abiotic stress-signaling pathways where calcium serves as a second messenger and particularly where regulation of ion transporter activities is involved. For example, low potassium stress, which presumably triggers a cytosolic calcium signal, activates CIPK23 via CBL1 and CBL9 to phosphorylate and activate the potassium channel AKT1. SCaBP1 interacts with and activates.

PKS5/CIPK11 and PKS24/CIPK14, which phosphorylate and inhibit $H^+$-ATPase on the plasma membrane. This inhibition is important for cellular pH regulation. CBL2 and CBL3 function redundantly with their interacting proteins CIPK3/9/23/26 to regulate $Mg^{2+}$ sequestration in the vacuole that is important for tolerance to high $Mg^{2+}$ stress. Other CBL-CIPK combinations are known to regulate various transporters important for plant responses to nitrate, ABA, or other abiotic stresses.

## Osmotic Stress Signaling

Plants have large families of MAP kinase pathway components, which can combine to form thousands of MAP kinase modules. For example, Arabidopsis contains over 60 MAP kinase kinase kinases (MAP3K), 10 MAP kinase kinases (MAP2K), and 20 MAP kinases (MAPK). The rapid activation of multiple MAPKs, including MAPK3, 4, and 6, has long been observed in plants in response to biotic as well as abiotic stimuli such as salt, drought, cold, heat, and wounding and in response to growth and developmental signals. The challenge in defining MAPK-signaling pathways for abiotic stress remains in the identification of upstream sensor proteins, in the identification of MAP3Ks and MAK2Ks that are responsible for the MAPK activation, and in connecting the kinase activation to downstream effector proteins and physiological outputs. It is still unclear whether a MAPK pathway similar to the yeast high osmolarity glycerol (HOG) MAPK pathway may function to mediate osmoregulation in plants. Perhaps the salt and drought-activated MAPKs are activated mainly by stress-triggered secondary signals rather than by the primary osmotic stress signal.

Salt, drought, and osmotic stress treatments also rapidly activate the SnRK2 family of protein kinases. In Arabidopsis, all 10 SnRK2s except SnRK2.9 are activated by osmotic stress, and SnRK2.2/3/6/7/8 are also activated by ABA. Although the mechanism by which ABA activates the SnRK2s has been elucidated, how osmotic stress activates the kinases is not known. Genetic evidence shows that tolerance to osmostic stress requires SnRK2s because decuple Arabidopsis mutant plants with all 10 SnRK2s disrupted are very sensitive to osmotic stress inhibition of growth. The decuple snrk2 mutant plants are impaired in osmotic stress regulation of a large number of genes and in the accumulation of ABA, the compatible osmolyte proline, and the second messenger $IP_3$, but they are unaffected in the osmotic stress-induced accumulation of ROS. These results suggest that SnRK2 activation is upstream of ABA accumulation and that this activation also controls osmotic adjustment and other adaptive responses to osmotic stress. Future efforts should be directed at finding upstream components important for SnRK2 activation under osmotic stress conditions and at identifying substrate proteins of osmotic stress-activated SnRK2s. Because osmotic stress triggers a cytosolic calcium signal and because the calcium channel OSCA1 is a putative osmosensor, candidate factors upstream of the SnRK2s include calcium-responsive kinases such as CPKs and SCaBP/CBL-PKS/CIPKs. In the moss Physcomitrella patens, recent work showed that a Raf-like kinase is critical for SnRK2 activation by both osmotic stress and ABA. It will be of interest to determine whether and how similar kinases may integrate osmotic stress and ABA signaling in higher plants.

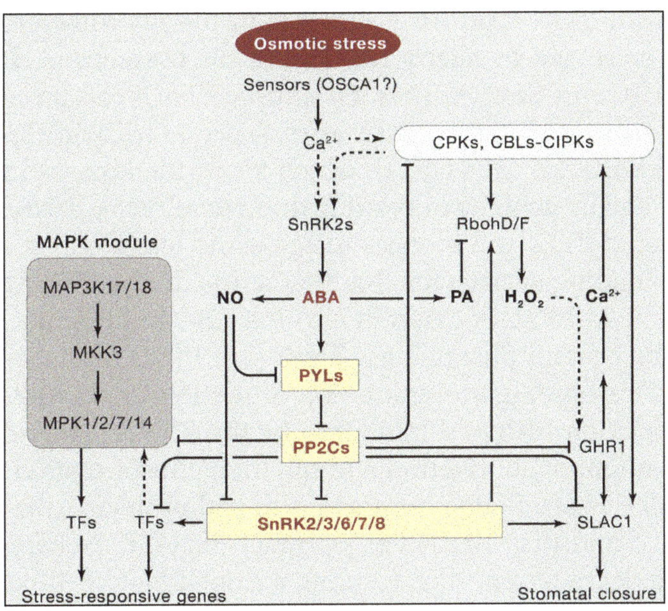


Osmotic Stress and ABA Sensing and Signaling The $Ca^{2+}$ channel OSCA1 may participate in osmosensing. The resulting $Ca^{2+}$ signal may activate CPKs and CBLs-CIPKs. Eventually, SnRK2s are activated, which leads to ABA accumulation. ABA binds to PYLs, which then interact with and inhibit class A PP2Cs, resulting in the activation of SnRK2.2/3/6/7/8. The activated SnRK2s phosphorylate effector proteins including

TFs (transcription factors), SLAC1, and RbohD/F. RbohD/F generates $H_2O_2$, which elicits a $Ca^{2+}$ signal through GHR1. This $Ca^{2+}$ signal activates CPKs and CBLsCIPKs that also phosphorylate effector proteins such as SLAC1. In addition to $Ca^{2+}$, ABA also induces the second messengers NO (nitric oxide) and PA (phosphatidic acid) and other phospholipids. NO inhibits SnRK2s and PYLs, and PA regulates proteins like Rbohs. Also depicted is ABA activation of a MAP kinase module. Components of the core ABA-signaling pathway are colored in red. Arrows indicate activation, bars indicate inhibition, and dashed lines indicate postulated regulation.

Osmotic as well as temperature stress cause the generation of various lipid signals including phosphatidic acid, phosphoinositides, sphingolipids, lysophospholipids, oxylipins, N-acylethanolamines, and many others. How stress regulates biosynthesis enzymes to generate the lipid signals is not well understood. In general, the lipid molecules bind to signaling proteins and affect their activity and membrane associations.

## ABA Signaling

The ABA-signaling pathway is central to drought and salt-stress responses in plants. One of the most important advances in stress signaling in the past decade has been the identification of ABA receptors and the elucidation of the core ABA-signaling pathway. Chemical genetics and protein-interaction studies led to the identification of the PYR/PYL/RCAR (hereafter referred to as PYL) family of soluble, START domain proteins as receptors for ABA. PYLs bind ABA with KD in the micromolar range, though the binding affinity is increased by nearly 100-fold in the presence of clade A PP2Cs, such as ABI1, ABI2, HAB1, and PP2CA. Therefore, these PP2Cs can be considered as co-receptors. In the absence of ABA, the PP2Cs are associated with SnRK2 kinases including SnRK2.2, 2.3, and SnRK2.6 (aka OST1), which keeps the kinases inactive by blocking their catalytic cleft and by dephosphorylating the activation loop. ABA enters the central hydrophobic pocket of PYLs and induces the gate and latch loops to close and lock the pocket, creating a binding surface for the PP2Cs. Inside the PYL-ABA-PP2C complex, a tryptophan residue in PP2C is inserted into the ABA-binding pocket and locks ABA tightly in place. The protein phosphatase activity of PP2C in the complex is inhibited by the ABA-PYL. This binding and inhibition of the PP2Cs by ABA-PYLs releases the SnRK2s from association with and inhibition by the PP2Cs. The released SnRK2s are activated through autophosphorylation and can then phosphorylate many downstream effectors. The small GTPase ROP11 interacts with and protects ABI1 from inhibition by PYL9. In turn, ABI1 and the other PP2Cs protect the GTP exchange factor RopGEF1 from ABA-induced degradation, thus forming a RopGEF-ROP-PP2C control loop that may counteract leaky ABA signaling by monomeric PYLs in the absence of stress.

Much redundancy exists in the functions of Arabidopsis PYLs, although each PYL may have unique biochemical properties and expression patterns. Knockout of PYL8 leads to ABA insensitivity in lateral root-growth recovery from stress inhibition. The role of PYL8 in regulating lateral root growth is independent of the core ABA-signaling pathway because PYL8 directly interacts with MYB77, and this ABA-induced interaction enhances MYB77-dependent transcription of auxin-responsive genes. Similarly, PYL6 interacts with MYC2, a key transcription factor in jasmonate responses and thus connects ABA and jasmonate pathways. Single-gene mutations in most other PYLs do not result in drastic ABA phenotypes. In contrast, pyr1prl1ply2pyl4 quadruple mutant plants are insensitive to ABA in germination and seedling growth, and pyr1pyl1pyl2prl4ply5pyl8 mutant plants are even more resistant to ABA in not only germination and seedling growth but also stomatal closure.

ABA strongly activates SnRK2.2, SnRK2.3, and SnRK2.6/OST1 and weakly activates SnRK2.7 and SnRK2.8. The snrk2.2/3/6 triple mutant in Arabidopsis is extremely insensitive to ABA in seed germination, seedling growth, stomatal closure, and gene regulation. Many effector proteins of ABA responses are direct substrates of SnRK2 kinases. bZIP transcription factors such as ABI5 and ABFs (ABA-responsive element-binding factors) are phosphorylated by the SnRK2s. Much of the ABA signaling occurs at the plasma membrane. The association of PYLs with the plasma membrane is mediated by their interaction with C2 domain proteins. Plasma-membrane proteins such as the anion channel SLAC1 are SnRK2 substrates that mediate ABA-induced stomatal closure and reduce transpirational water loss under drought stress. Recent phosphoproteomic studies identified dozens of additional SnRK2 substrate proteins, including several proteins important for chloroplast function, flowering time control, miRNA and chromatin regulation, and RNA splicing. The PYL-PP2C-SnRK2 core ABA-signaling module activates a MAPK cascade comprised of the MAP3Ks MAP3K17/18, the MAP2K MKK3, and the MAPKs MPK1/2/7/14, which may regulate many ABA effector proteins through phosphorylation.

ABA-activated SnRK2s also phosphorylate the plasma-membrane NADPH oxidase RbohF, which when phosphorylated generates $O_2^-$ in the apoplastic space. The $O_2^-$ subsequently forms $H_2O_2$, a signaling molecule that mediates various ABA responses, including stomatal closure. ABA induction of ROS in guard cells is impaired in Arabidopsis pip2;1 mutant plants, indicating that apoplastic $H_2O_2$ can enter cells via the aquaporin PIP2;1. The MAP kinases MPK9 and MPK12 function redundantly in anion-channel regulation in guard cells downstream of ROS and affect ABA regulation of stomatal closure. Another important component that connects ABA signaling and ROS is the plasma-membrane RLK GHR1.

GHR1 interacts with and activates SLAC1. GHR1 is critical for ABA and ROS regulation of stomatal closure. Interestingly, GHR1 function is antagonized by ABI2 but not by ABI1. $H_2O_2$ could also modulate the calcium signal to affect ABA responses, and GHR1 is required for $H_2O_2$ activation of plasma-membrane $Ca^{2+}$ channel. Calcium signaling is critical for ABA regulation of stomatal closure, and mutant plants defective in four redundant calcium-dependent protein kinases, CPK5, CPK6, CPK11, and CPK23, are incapable of closing stomata in response to ABA. Like the SnRK2s, CPKs can phosphorylate effectors (including SLAC1) of ABA responses in guard cells. Furthermore, ABA-triggered calcium signal can activate the CBL1/9-CIPK26 module to cause phosphorylation of effector proteins such as RbohF. In addition to inducing $H_2O_2$ and calcium signals, ABA triggers the generation of nitric oxide (NO) and phospholipids such as phosphatidic acid. NO causes the S-nitrosylation of a cysteine residue adjacent to the catalytic site of SnRK2s, resulting in inactivation of the kinases. NO also causes tyrosine nitration and S-nitrosylation of cysteine residues in PYLs. Tyrosine nitration inhibits PYL activity and also accompanies polyubiquitylation and proteasome-mediated degradation of PYLs. Phosphatidic acid contributes to ABA signaling by binding to and activating RbohD and RbohF. Therefore, the in vivo regulation of SLAC1, Rbohs, and other effectors of ABA responses in plants requires a network of signaling pathways that include not only the PYL-PP2C-SnRK2 core pathway but also other pathways that involve calcium, ROS, NO, phospholipids, and the other kinases.

## Cold and Heat Stress Signaling

Cold stress greatly affects plant metabolism and transcriptomes. The effect on plant metabolism arises from both direct inhibition of metabolic enzymes by cold temperatures and reprogramming of gene expression. In temperate plants, an exposure to low, non-freezing temperatures enhances the tolerance to subsequent freezing temperatures, a process known as cold acclimation. Cold stress rapidly induces the expression of many transcription factors, including the AP2-domain proteins CBFs, which then activate the expression of numerous downstream cold-responsive (COR) genes. The CBF genes are controlled by upstream transcription factors, including the bHLH transcription factor ICE1. ICE1 is subjected to sumoylation and polyubiquitylation and subsequent proteasomal degradation, mediated by the SUMO E3 ligase SIZ1 and ubiquitin E3 ligase HOS1, respectively. Cold-stress induction of the CBF and COR genes is gated by the circadian clock, whose activity is modulated by diurnal oscillation of the plastid retrograde signal tetrapyrroles.

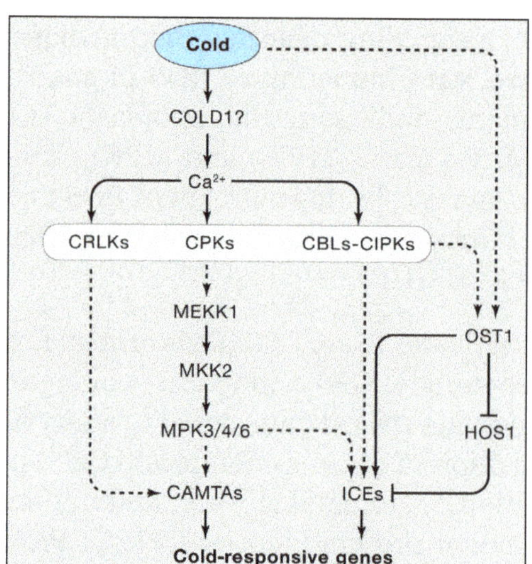


Cold-Stress Sensing and Signaling.

Ding et al. recently reported that cold stress activates SnRK2.6/OST1 and that SnRK2.6 interacts with and phosphorylates ICE1 to activate the CBF-COR gene-expression cascade and freezing tolerance. Furthermore, cold-activated SnRK2.6 disrupts the interaction between ICE1 and HOS1 to prevent ICE1 degradation. The reported cold-stress activation of SnRK2.6 is independent of ABA, although the activation is negatively regulated by ABI1 and other PP2Cs. Therefore, the activation presumably does not involve PYL ABA receptors. There is discrepancy in the reported cold activation of SnRK2.6, given that Boudsocq et al. reported that none of the 10 Arabidopsis SnRK2s is activated by cold stress.

Cold stress is sensed by membrane proteins such as COLD1, leading to a cytosolic $Ca^{2+}$ signal. CPKs and CBLs-CIPKs may mediate the $Ca^{2+}$ signal to activate a MAP kinase cascade. The activated MPKs are postulated to phosphorylate TFs such as CAMTAs and ICE1/2, which then activate cold responsive genes. Through an unknown mechanism, cold stress also activates OST1 (SnRK2.6), which inhibits HOS1, and phosphorylates and activates ICE1. Arrows indicate activation, bars indicate inhibition, and dashed lines indicate postulated regulation.

Teige et al. showed that the Arabidopsis MAP2K MKK2 is activated by cold and salt stress and controls COR gene expression and plant tolerance to freezing and salt stress. MKK2 is downstream of the MAP3K MEKK1 and upstream of MPK4 and MPK6, which are activated by various stresses including cold temperatures. Pharmacological studies indicate that membrane fluidity and cytoskeleton and calcium influx are involved in cold-stress regulation of COR genes and MAPKs. The receptor-like kinase CRLK1 may connect cold-stress-induced calcium signaling with the MAPK cascade because CRLK1 binds to calcium and calmodulin, interacts with MEKK1, and is requisite for cold-stress activation of MAPK activities. It appears that cold stress affects membrane fluidity, which may be sensed by plasma-membrane proteins such as calcium channels or associated proteins, resulting in calcium influx and activation of calcium-responsive protein kinases (CPKs CIPKs, and CRLK1) and the MAPK cascade, which regulate COR gene expression. In the future, it will be necessary to use genetic, molecular, and biochemical analyses to clarify the roles of these kinases and the relationships among them as well as between them and SnRK2.6 in cold-stress signaling.

Heat stress induces the expression of HSPs, many of which function as molecular chaperones to prevent protein denaturation and maintain protein homeostasis. Like mammalian heat stress transcription factors (HSFs), plant HSFs are released from association with and inhibition by the HSP70 and HSP90 chaperones as a result of the binding of the chaperones to misfolded proteins caused by heat stress, and thus the HSFs are made available to activate heat stress responses. Heat stress also activates MAPKs, which modulate HSP gene expression. The MAPK activation may be linked to heat-induced changes in membrane fluidity and calcium signaling that are important for HSP gene expression and thermotolerance. Common features between cold and heat-stress signaling are not limited to membrane fluidity changes, calcium signaling, and MAPK activation but also include the involvement of ROS, NO, phospholipid signals, protein sumoylation, and proteosomal degradation.

## Systemic Signaling

Pathogen infection and wounding trigger systemic responses in plants. Similarly, abiotic stresses such as drought, salt, cold, heat, and high light also elicit systemic responses such that locally applied stress causes responses not only locally but also in distal tissues, resulting in systemic acquired acclimation (SAA). SAA involves long-distance electrical and hydraulic signals as well as calcium and ROS waves. Stress-triggered calcium and ROS waves can move at speeds that exceed 1,000 mm per second as visualized in transgenic plants expressing calcium-sensitive fluorescent protein and luciferase reporter driven by a ROS-responsive promoter, respectively. Calcium and ROS waves have been shown to cause transcriptional responses in distal tissues. ROS waves require the plasma-membrane NADPH oxidase RbohD, whereas calcium waves depend on the vacuolar ion channel TPC1, which may be involved in calcium-induced calcium release. RbohD is phosphorylated by the calcium-dependent protein kinase CPK5, which can be activated by ROS and is required for systemic defense responses. A model that emerges is that ROS generated by NADPH oxidases triggers a cytosolic calcium signal to activate more NADPH oxidases via calcium responsive kinases, thus generating a self-propagating mutual activation circuit between ROS and calcium signals. $H_2O_2$ generated by RbohD may enter cells through a plasma-membrane-intrinsic protein/water channel (PIP), and its activation of the plasma-membrane calcium channel likely requires GHR1 or a similar RLK. The plasma-membrane calcium channel that is activated by ROS has not been identified. In addition, it is unclear how the calcium and ROS waves may be linked to long-distance electrical signals.

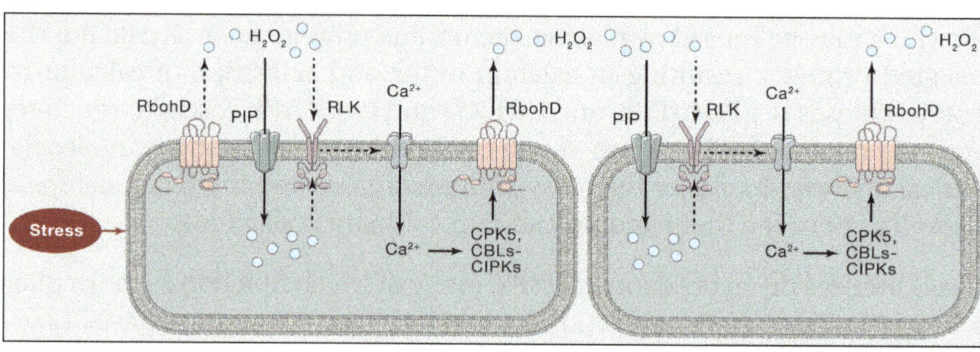


Model of Systemic Stress Signaling Local exposure to stress generates $H_2O_2$ and $Ca^{2+}$ signals. The $Ca^{2+}$ signal can activate CPKs and CBLs-CIPKs, which phosphorylate and activate RbohD. Activated RbohD generates $H_2O_2$ that diffuses through the cell wall to neighboring cells, where it induces a $Ca^{2+}$ signal through RLKs like GHR1. $H_2O_2$ may activate $Ca^{2+}$ signaling at the cell surface and may also enter the cell through PIP water channels and then activate $Ca^{2+}$ signaling intracellularly. The mutual activation between the $Ca^{2+}$ and $H_2O_2$ signals generates self-propagating calcium and ROS waves that can travel to distant tissues to cause systemic acquired acclimation responses. Dashed lines indicate postulated regulation.

## References


- Abiotic-stress, medicine-and-dentistry: sciencedirect.com, Retrieved 13 June, 2020
- Savvides, Andreas (December 15, 2015). "Chemical Priming of Plants Against Multiple Abiotic Stresses: Mission Possible?". Trends in Plant Science. 21 (4): 329–340. doi:10.1016/j.tplants.2015.11.003. hdl:10754/596020. Retrieved March 10, 2016
- Abiotic-stress-in-harvested-fruits-and-vegetables, abiotic-stress-in-plants-mechanisms-and-adaptations: intechopen.com, Retrieved 29 July, 2020
- Gao, Ji-Ping; et al. (2007). "Understanding Abiotic Stress Tolerance Mechanisms: Recent Studies on Stress Response in Rice". Journal of Integrative Plant Biology. 49 (6): 742–750. doi:10.1111/j.1744-7909.2007.00495.x
- Abiotic-and-biotic-stress-in-plants/the-effect-of-climate-change-on-abiotic-plant-stress-a-review: intechopen.com, Retrieved 11 March, 2020
- Wang, W.; Vinocur, B.; Altman, A. (2007). "Plant responses to drought, salinity and extreme temperatures towards genetic engineering for stress tolerance". Planta. 218 (1): 1–14. doi:10.1007/s00425-003-1105-5. PMID 14513379

# Types of Biotic and Abiotic Stress

CHAPTER 3

There are many types of biotic and abiotic stress such as chilling stress, heat stress, moisture stress, flood stress, drought stress, mineral stress, oxidative stress, ultraviolet radiation stress, salinity stress, etc. The topics elaborated in this chapter will help in gaining a better perspective of these types of biotic and abiotic stress in plants.

## CHILLING STRESS

Chilling temperatures effects on plants in temperate climates lead to a reduction or complete crop failuredue to either direct damage or delayed maturation. Evena small drop in temperature, causing no visible damage tochilling-sensitive plants, caused to up to 50% reduction in their productivity. For example, chilling damage to young cotton plants in U.S. in 1980 resulted in the loss of 60 million dollars. In South and South-East Asia, high-yielding varieties of rice are not grown in areas of more than 7 million hectares, where they may be exposed to chilling temperatures. Obviously, the problem of plant resistance to chilling temperatures, which often occur in spring and autumn in many countries, is important for practical plant breeding.

Table: The list of the vegetables, sensitive to chilling temperatures, the lowest safestorage/handling temperature andthe symptoms of chilling injury.

| Crop | Lowest safe temperature °C | Chilling injury symptoms |
|---|---|---|
| Asparagus | 0–2 | dull, gray-green, limp tips |
| Bean (snap) | 7 | pitting and russeting |
| Cucumber | 7 | pitting, water-soaked lesions, decay |
| Eggplant | 7 | surface scald, Alternaria rot, seed blackening |
| Okra | 7 | discoloration, water-soaked areas, pitting, decay |
| Pepper | 7 | pitting, Alternaria rot, seed blackening |
| Potato | 2 | mahogany browning, sweetening |
| Pumpkin | 10 | decay, especially Alternaria rot |
| Squash | 10 | decay, especially Alternaria rot |

| | | |
|---|---|---|
| Sweet potato | 10 | decay, pitting, internal discoloration |
| Tomato (ripe) | 7–10 | water-soaking, softening, decay |
| Tomato (mature-green) | 13 | poor colour when ripe, Alternaria rot |

The most noticeable visual symptoms of chilling injury in herbaceous plants are leaf and hypocotyls wilting,which often precedes the appearance of infiltration (water saturated areas), the appearance of surface pits and large cavities, discoloration of leaves and internal tissues, accelerated aging and rupture of chilled tissues, slow, incomplete or uneven ripening, accompanied by a deterioration of the structure and flavor; increased susceptibility to decay, drying of the edges or tips of leaf blades and in the caseof prolonged chilling – leaf necrosis and plant death. According to Skog, potentialsymptoms of chilling injury are surface lesions, water soaking of tissues, water loss, desiccation or shrivelling, internal discolouration, tissue breakdown, failure of fruit to ripen, or uneven or slow ripening, accelerated senescenceand ethylene production, shortened storage or shelf life, compositional changes, loss of growth or sprouting capability, wilting and increased decay due to leakageof plant metabolites, which encourage growth of micro organisms,especially fungi.

A characteristic effect of chilling temperature son chilling-sensitive plants is growth slowing, more pronounced in susceptible species and varieties in comparison with the tolerant species. In addition, there is a delayed development and lengthening of the growing season. At the same time apical cone differentiation is delayed, reducing thenumber of newly formed plant organs and the rate oftheir occurrence, the structure of roots is changed, and flowering rate, fruit and seed filling are reduced.

## Effect of Chilling on the Physiological Processes in Chilling-sensitive Plants

Incubation of chilling-sensitive plants at low temperatures induces disturbances in physiological processes: Water regime, mineral nutrition, photosynthesis, respiration and total metabolism.

Water regime: Chilling of sensitive plants affectsall components of water regime and causes loss of water, resulting in strong wilting. It is based on thetwo main factors: rapid decline in the ability of roots toabsorb water and transport it to the shoots and reduced ability to close stomata in response tosubsequent water deficit. Insufficient water supplyprovokes rapid drop in water potential of leaves duringthe first hours of cooling. The degree of chilling damage of plants can be reduced by means of preventing the disturbance of the water regime.

Mineral nutrition: Low temperatures have aneffect on mineral nutrition of plants. Absorption of ions by roots is difficult, as well as their movement in the above ground parts of plants. The distribution of nutrients between the plant organs is disrupted, with general decrease in the nutrient content in the plant.

Chilling of plants leads to a decrease in the activity of nitrate reductase, reduction in the nitrogen incorporation inthe amino acids and proteins, and a drop in the proportion of organic phosphorus and an increase in inorganic P content, which is a consequence of a breach of phosphorylation and enhanced decomposition of organic Pcompounds. Mechanisms to reduce the absorption of nutrients by chilling temperatures include depression of respiration and oxidative phosphorylation,impair enzymatic transport systems associated with conformational proteins changes in membranes,changes in membrane potential, reducing the supply ofATP to $H^+$-transporting ATPase, as well as lowering the permeability coefficients for ions.

Respiratory rate: The consequence of keeping plants at chilling temperatures is a change in respiratoryrate. There is evidence of its decline, occurring as a result of destruction of the mitochondria structure, thegeneral lowering of kinetic energy, and the inhibition of some enzymes. The mechanism of stimulation is unknown, but it is possibleto assume that it was the result of uncoupling of oxidative phosphorylation. It is also possible that the increased respiration reflects a reaction to the transfer ofplants from chilling temperatures to the higher temperatures. As a result of decreased respiration and increased consumption of energy-rich phosphates at chilling temperatures is a reduction of ATP levels.

Cold-tolerant crop species have greater temperature homeostasis of leaf respiration than cold-sensitive species. Chilling reduces the cytochrome path of the electron transport in seedlings and enhances alternative.

Respiratory pathways: Perhaps these alternative pathways play animportant role in plant adaptation to chilling. They are triggered at the chilling period and increase with decreasing temperature. These alternative pathways induced by chillingcaused a decrease in superoxide generated in mitochondria.

Rate of photosynthesis: During and after chilling, the rate of photosynthesis in the leaves of chillingsensitiveplants decreased and this is more related todecreasing temperature and lengthening of chilling periodand persisted for a long time after transfer of chilledplants in the heat. The physiological reasons for the suppression ofphotosynthesis are the inhibition of phloem transport of carbohydrates from the leaves, stomatal limitation, destruction of the photosynthetic apparatus, damage to water-splitting complex of photosystem I, inhibiting electron transport, and uncoupling of electron transfer andenergy storage, changes in the activity and inhibition of synthesis of key enzymes of the Calvin cycle and $C_4$-way. Cold-sensitive crop species have smaller temperaturehomeostasis of leaf photosynthesis than cold-tolerantspecies.

## The Theory of Chilling Injury

In the initial period of studying, the influenceof low temperatures on chilling-sensitive plants waswidespread theory Sachs about the death of plants dueto disorders of water regime. However, subsequent studieshave shown one-sided interpretation of these data.

Changes in water regime were likely due to disturbances of other processes. In the middle of the 20th century it was found that the wilting of the aerial organs is not dueto excessive transpiration over slow supply of water by roots, but is the result of lowering water-holding capacity due to disorganization of the cytoplasm structure and metabolic decompensation.

Based on observations of changes in protoplasmic viscosity at low temperatures, it has been suggested that this cell property plays a key role in the damage. The less tolerant plants to cold, the higher temperature at which cytoplasm geling occurs andthe faster increases the viscosity of the cytoplasm. At considerable increase in viscosity the rate of biochemical reactions in the cytoplasm is decreased, the metabolismis disturbed, which leads to dysfunction of physiological processes. However, it was shown that cucumber plants decreased viscosity with increasing chilling duration, and the worst after 2.5–4 days, and then increased gradually, reaching viscosity level of non-chilled plant, but shortlybefore the complete withering away could exceed thislevel. An increase in viscosity of highly damaged plantsalso continued after the transfer into heat. "Dying" increase protoplasmic viscosity and is the final stage of death, which has no relation to the first stage of damage, but only deepens it.

Among the hypotheses about the primary events that trigger the occurrence of reaction to lower temperatures, hypothesis of phase change prevailed in the 1970s,according to which the hilling-induced changes in themolecular ordering of membrane lipids may be the causeof chilling injury. These changes increase the damage by lowering the ATP levels, metabolic imbalances and increasing membrane perme ability. However, all these changes do not occur immediately after the start of chilling and are likely to be secondary disorders. The increase in membrane permeability due to the low-temperature exposure ("membranes leakage") should be quick, registered in the first few minutes after placing the tissue at chilling temperatures, in accordance with the hypothesis of phase transitions. In reality, this does not happen, and often passive permeability is not increased.

Moreover, the increase in electrolyte leakage is likely dueto chilling-induced water stress, which has been revealedto be considerably weaker in an atmosphere saturated with water. At the same time it is note worthy that the low un saturation of membranous phospholipides, which is generically determined, gives sensitivity to cold temperatures to chilling-sensitive plants. The data about the introduction of genes offatty acid desaturases in a genome of chilling-senitiveplants confirms that this gives sensitive plants more pronounced chilling resistance.

In recent years, special attention of researchershas been drawn to two hypotheses to explain the induction of chilling damage to a rapid increase in the concentration of free cytosolic $Ca^{2+}$ ($[Ca^{2+}]_{cyt}$) and the occurrence of oxidative stress upon chilling of chilling-sensitive plants. Minorsky proposed a hypothesis to explain most of the secondary effects of chilling shock, which suddenly increases (by 1–2 orders) in the concentration of $[Ca^{2+}]_{cyt}$. It is assumed that the rapid increase in $[Ca^{2+}]_{cyt}$ due to chilling, may serve as the primary physiological signal of cold exposure. It was shown that changesin intracellular calcium compartmentation in chilled plants, leading to an increase in $[Ca^{2+}]_{cyt}$, stop cytoplasmic streaming and affect the subcellular structures. There is evidence that input of $^{45}Ca^{2+}$ inmaize root cells increased by 20–25% at a temperature of 2°C. Changes in $[Ca^{2+}]_{cyt}$ trigger cascade reactions in the cell, which leads to numerous disturbances at all levels of an organization. Our investigation shows that chilling induces abrupt reduction of $Ca^{2+}$-ATPase activity, which pumps out $Ca^{2+}$ in apoplast and in intracellular depots.So, this enhances the $[Ca^{2+}]_{cyt}$ level in cytoplasm. During the growth of maize seedlings on nutrient media with different calcium status more intense chilling injury was observed at reduced or enhanced $Ca^{2+}$ doses in comparis on with optimal dose.In recent years, the calcium hypothesis has been further developed in view of oxidative stress that occurs when cooling the chilling-sensitive plants. Oxidative stress that occurs during cooling of chilling-sensitive plants plays a leading role in the transduction of chilling injury. The reason why production of free radicals and reactive oxygen species (ROS) increased is singlet oxygen, superoxide anion,hydroxyl radical, hydrogen peroxide. These ROS cause considerable damage to membrane lipids and other cellular components. It was shown that $[Ca^{2+}]_{cyt}$ changes are intimately connected to an oxidative stress. Oxidative stress causes an immediate increase in cytosolic calcium, acting the same as chilling shock.

This reaction is transient, and finishes within 1–2 minutes. Inturn, $[Ca^{2+}]_{cyt}$ influences a level of free radicals, inhibiting activity of SOD. So, increasing the concentration of ionized calcium causes increased oxidative stress, i.e. is the signal amplification cascade that causes chilling damage.

## Ways to Improve Chilling Tolerance of Chilling-sensitive Plants

At the present time, to improve the chilling tolerance of sensitive plants various techniques are used, which can be divided into several groups: The thermal effect, chemical treatment, the use of cellular and genetic engineering.

Thermal effects includes low-temperature hardening, thermal conditioning, intermediate arming, andthe effect of heat stress. The basis of seed and seedling hardening of chilling-sensitive plants to cold, which haslong been used in practical plant breeding, is the adaptation of the organism in the early stages of development, accompanied by the emergence of specific structural and functional re arrangements. Low-temperature hardening process is associated with the protein-synthesizing system and is accompanied by a re arrangement of the hormonal system of plants.

Intermediate warming is another way of thermal regulation of chilling injury. Transfer of the chilled plants in the warm afternoon prevented the appearance of visible symptoms of damage, impaired inhibition of photosynthesis and transpiration, reduced leaf osmotic potential. Intermediate warming is often used for storage of chilling-sensitiveplants' fruits. It is assumed that the temporary placement in heat allows the chilled tissues to metabolize toxic substances that accumulate during the chilling process, or helps to restore the compounds in tissues that are depleted during chilling.

Cytokinins and ABA were most effective of allplant growth regulators. Non-hormonal growth regulators are used also in orderto improve the chilling tolerance of cultivated plants.These include paklobutrazol, chlorocholin chloride, me fluidid, unikonazol and other triazoles. The treatment by antioxidants and freeradicals quenching (ethoxyquin, sodium benzoate,glutathione, tyrone, formate, ascorbate, diphenylamine,α-tocopherol, propyl gallate) can slow down the degradation of unsaturated fatty acids and reduce chilling damagein chilling-sensitive plants, leaves and fruits.

Cellular and genetic engineering is a new trend, which allows fundamental changes in the chilling resistance of chilling-sensitive plants. They are based on a large genetic variability in components, controlling sensitivity, on the one hand, and on the development of gene transfer technology, transformation and selection markers, on the other hand. So, screening the surviving cells during chilling of callus and suspension cultures and subsequent plant regeneration yielded plants with increased epigenetic resistance to chilling temperatures.

## HEAT STRESS

The sudden increase in ambient maximum temperature, in a matter of few days, by 5–7°C with corresponding increase in the minimum temperature, creates "heat stress" on plants. The normal physiology of the plant gets affected and plant maturity is accelerated. In some cases, plants shed leaf/flower/fruit or dry-up non productive tillers and even become sterile to overcome the unprecedented stress. In practical agriculture, such heat stress inflicts enormous crop losses. Due to global weather change, the frequency of heat stress is predicted to increase in different parts of the world.

Rates of photosynthesis and respiration increase with an increase in temperature until a threshold maximum photosynthesis level is achieved. Generally, it is around 22–24°C and remains that way, up to 30–32°C. In rice, it was found that grain yields declined if mean seasonal temperature increased above 33°C. Surpassing the high temperature peak, various enzymes get inactivated, decreasing the photosynthetic efficiency. The high temperature coupled with high respiration and evaporation pushes the plant to permanent wilting when the temperature exceeds 46°C. The extent of crop loss inflicted by heat stress can ruin the income for poor farmers in heat prone areas. In many cases, there may be adequate soil moisture butthe negative water balance created by heat stress lead to withering of plants in a matter of few days. Under the natural habitat, plants have been exposed to more than one stress at a time; consequently they have evolved strategies against a combination of stresses by responding to them in different ways. Often high temperature, high transpiration rate and high light intensity occur together affecting the plant productivity.

Short periods of very high temperature (>35°C) are of common occurrence in many wheat-growing areas of the world. Such sudden exposure to heat stress affects the yield and quality of the wheat grain. Maturing kernel is highly sensitive to heat stress at milk and dough stages of grain development. The kernel becomes progressively less sensitive as the grain hardens. Reduction in kernel weight results primarily due to the shortening of the grain filling duration than due to the reduction in the rate of grain filling. The grain filling is seriously impaired and translocation processes get affected due to early senescence of the leaf and ear, and the reduction in chlorophyll content adversely affects photosynthesis. Since the photosynthetic source is affected and the sink duration is reduced, filling of the kernel is poor, resulting in small grain size having less grain weight. Under such situations, alternate sugar sources, stored as stem reserves, get mobilized so as to fill the kernel. It has been well demonstrated in spring wheat cultivars that comparable genotypes differ in grain shriveling under drought and heat stress conditions. This genotypic variation in the mobilization of stored stem reserves to support grain filling is an important consideration in selecting wheat lines for heat tolerance. Also, it is widely acknowledged that in wheat, under heat stress situations, kernel number per spike is a reliable measure for heat tolerance.

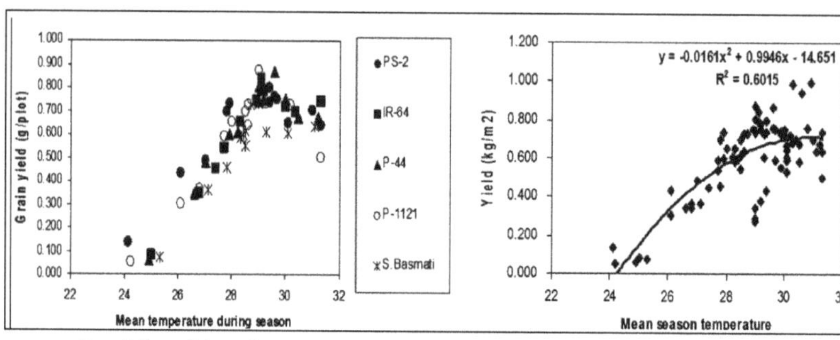


Realtionship of mean season temperature with grain yield of rice cultivars. Means seasons temperature 33°c and below 27°c drastically reduced rice yiled both in aromatic and non-aromatic cultivars.

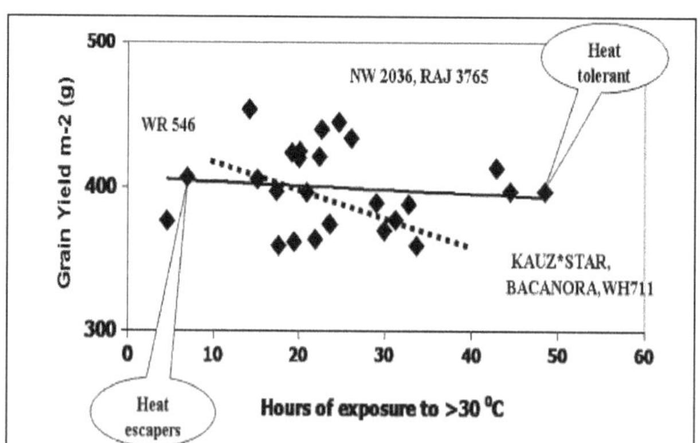


Effect of high tempreture on grain yield in diverse bread wheat genotypes. Twenty-five genotypes were fiels grown in six location and were exposed to different temperature regimes.

## D-impact on Quality of the Harvest

High temperature during seed development is associated with a reduction in total oil yield. Under field conditions the negative effect of high temperature gets amplified if there is also a moisture stress. Both temperature and moisture stress together influence the yield of oil in Brassica by interfering with seed growth and development. At elevated temperatures, with high night temperatures, a marked reduction in the percentage of linoleic acid occurs, apparently due to desaturase enzymes that are essential for the conversion of oleic to linoleic acid. It is now well recognized that reduced yields and altered oil composition occur in sunflower crop that matures under high temperature conditions. Whole ranges of metabolic changes that occur due to abnormal temperature cycle adversely affect the biosynthesis of fatty acids in sunflower. High maximum temperature negatively influences the yield of spring wheat affecting the number of productive tillers per square meter, grain weight, grains per spike and quality of the grain protein. And in cotton, both yield and fiber quality gets affected by noncardinal temperature regimes. In aromatic and non-aromatic rice cultivars, elevated temperatures, during grain filling period, decrease the amylase content of the grain.

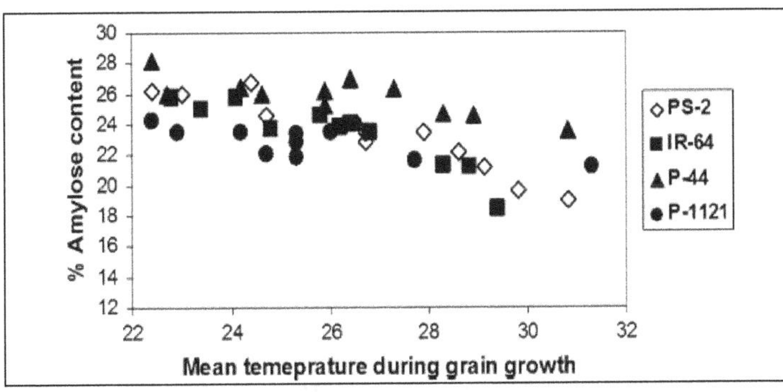


Effect of mean temperature, during grain-growth period, on amylase in rice cultivars. Amylase content decreased at the rate of 0.44% per degree rise in temperature in the cultivars.

## Physiological Responses to High Temperature

## High Temperature Limits to Optimal Plant Performance

Plants can experience wide fluctuations of temperature on a daily or seasonal basis. It is to be expected, therefore, that plants have evolved mechanisms through which cellular activity could be maintained at varying temperatures. These mechanisms have not been fully defined. It has been estimated that the optimal thermal range for any plant is about 108 °C wide. Exposure to temperatures outside this optimal range, though not necessarily lethal, can be considered stressful. The actual upper temperature limits for survival, which for temperate plant species is between 408 °C and 558 °C, depending on the length of exposure, act to limit plant distribution and the area available for agriculture.

Plants show differential sensitivity to high temperature depending on the severity, duration and developmental timing of the stress. This makes it difficult to estimate the damage done to crops by heat stress. The connection between plant water status and temperature also complicates analysis of the relative of contribution of heat and drought stress under field conditions. However, for maize, the best single variable for estimating yield was reported to be the extent to which the daily maximum temperature exceeded 328 °C during pollination and seed fill. Similarly, the highest yields of soybean have been correlated with the coolest reproductive season temperatures, and optimal temperature for high yield in cereal crops has been determined to be 20–308 °C. Not only moderate midday heat, but also high temperatures at night are damaging, leading to increases in carbon loss by respiration, decreased yield, alterations in timing of flowering, and even inhibition of flower bud development. These relatively mild high temperatures contrast with situations in which plants are challenged by severe stress temperatures, such as in the semiarid tropics where cereal grains can experience soil surface temperatures of 558 °C. In closed canopy maize fields in the midwestern USA leaf temperatures can be over 408 °C, while developing pea seeds in Idaho have been recorded to have temperatures up to 688 °C. Altogether, it seems likely that many plants experience some form of heat stress during their life cycle.

## Heat Sensitivity of Photosynthesis

The fact that heat stress limits photosynthetic output is well documented, and photosynthesis declines at temperatures well below those lethal to the plant. However, the mechanism by which this occurs remains controversial. Current data preclude distinguishing whether the decline in photosynthesis at elevated temperatures occurs by a common mechanism or a diversity of mechanisms. Numerous components of the photosynthetic apparatus display heat lability, but it has been difficult to determine which one is rate limiting under conditions of heat stress. It has long been proposed that PSII is the most heat-sensitive component of the photosynthetic apparatus. Other parameters implicated as weak links during heat stress include components that facilitate $CO_2$ transfer from the intercellular space to the chloroplast, photosynthetic electron transport and loss of ribulose bisphosphate (RuBP) regeneration capacity.

Recently, Salvucci and Crafts-Brandner have argued that the heat lability of various photosynthetic components is secondary to the failure of cells to maintain RuBP carboxylase/oxygenase (Rubisco) in an optimally activated state.Under conditions of heat stressRubisco activity declines as a consequence of the association of catalytic 'misfire products' at the active site rather than temperatureinduced enzyme inactivation. It is the well-characterized temperature sensitivity of Rubsico activase that precludes adequate reactivation of these dead-end complexes during periods of heat stress. Interestingly, in wheat and cotton leaves specific isoforms of activase accumulate during heat stress, which may add stability to the multisubunit chaperone-like holoenzyme. Arguing against a role for Rubisco activase as a limiting factor in photosynthetic output during heat stress, Wise et al. report that Rubisco activity in field-grown cotton remains sufficient during heat stress to process levels of RuBP supplied to the enzyme, and hence maintain upstream photosynthetic processes. Clearly, more work will be required to sort out the factors that limit photosynthetic output during heat stress. Finally, it should be noted that an alternate function for Rubisco activase was suggested by Rokka et al. as a consequence of their finding that the enzyme was reported to relocate to thylakoid membranes under conditions of heat stress, where it appears to associate with thylakoidbound polysomes. Whether it acts as a chaperone at this location has not yet been explored.

Interest in the consequences of global warming has led to analysis of the influence of $CO_2$ levels on the heat sensitivity of photosynthesis, with both negative and positive consequences documented. For example, the activation state of Rubisco declines with rising $CO_2$ concentrations, with the elevated $CO_2$ stimulating increases in ATP consumption that reduce ATP/ADP ratios, and thus possibly inhibit Rubisco activase. This effect would compound photosynthetic losses caused by the heat sensitivity of the enzyme. In contrast, a protective role for $CO_2$ was reported by Taub et al. Their results showed that leaves of Cucumis sativus and several other species grown in an atmosphere enriched to 750 ppm $CO_2$ retained significantly higher photosynthetic rates (measured as PSII activity or $CO_2$ uptake) after a heat shock at 408C. The heat treatments in these experiments were performed in the dark, thus minimizing the possibly confounding effects of photoinhibition.

## Heat Sensitivity of Reproduction

In addition to the studies cited above concerning effects of chronic high temperature on crop yield, numerous studies document that successful fertilization and seed fill are more sensitive to high temperature than vegetative growth. Unfortunately, reduced fertility and seed fill at high temperature cannot be clearly pinpointed to a defect in a single function, but most likely varies with the type of heat stress and the plant species. Loss of fertility could result from problems in male meiosis, pollen germination, pollen tube growth, or megagametophyte defects among other factors. Likewise, effects on flower production, grain set, endosperm division, source photosynthesis, and assimilate transport and partitioning can all contribute to ultimate seed yield and weight. Even recent studies are still only trying to define the heat-sensitive processes in different plant species.

Perhaps the most progress at the molecular level has been made in an examination of potential factors limiting starch production in wheat grain at high temperatures. Hurkman et al. observed transcript levels of starch biosynthetic enzymes comparing 24/178°C to 37/288°C and 37/178 °C stress regimes. Although transcript levels of three isoforms of starch synthase were dramatically lowered by the heat treatment, this effect was not mirrored in the rate of starch accumulation. The overall time of grain-fill was severely shortened, however, and the type of starch granule was altered. Heat effects on wheat starch synthase were also reported by Zahedi et al. who concluded that temperature differences in starch synthase efficiency were correlated with differences in the temperature sensitivity of grain fill between two cultivars. However, the complexity of the reproductive process is obviously consistent with multiple gene effects.

In nature, different plant species, populations and even individuals may employ very different strategies to optimize growth and reproduction in the face of high temperature stress. Plants have evolved morphological, life history, physiological and cellular strategies not only to cope with temperature, but also to avoid high temperature damage.

To date, the most successful strategies for enhancing agriculture in high temperature environments has primarily involved avoidance mechanisms, most often altering the timing of reproduction to achieve yield gains in hot climates. Further advances in our understanding of responses to heat and our ability to manipulate heat tolerance will greatly benefit from application of genomics and proteomics techniques coupled with additional genetic analysis.

## MOISTURE STRESS

Moisture stress occurs when the water in a plant's cells is reduced to less than normal levels. This can occur because of a lack of water in the plant's root zone, higher rates of transpiration than the rate of moisture uptake by the roots, for example, because of an inability to absorb water due to a high salt content in the soil water or loss of roots due to transplantation. Moisture stress is more strongly related to water potential than it is to water content.

Moisture stress also has an effect on stomatal openings of a plant, mainly causing a closure in stomata as to reduce the amount of carbon dioxide assimilation. Closing of the stomata also slows the rate of transpiration, which limits water loss and helps to prevent the wilting effects of moisture stress. This closing can be trigged by the roots sensing dry soil and in response producing the hormone ABA which when transported up the xylem into the leaves will reduce stomatal conductance and wall extensibility of growing cells. This lowers the rates of transpiration, photosynthesis and leaf expansion. ABA also increases the loosening of growing root cell walls and in turn increases root growth in an effort to find water in the soil.

Phenotypic response of plants to long-term water stress was measured in corn and showed that plants respond to water stress with both an increase in root growth both laterally and vertically. In all Droughted conditions the corn showed decrease in plant height and yield due to the decrease in water availability.

Genes induced during water-stress conditions are thought to function not only in protecting cells from water deficit by the production of important metabolic proteins but also in the regulation of genes for signal transduction in the water-stress response. There are four pathways that have been described that show the plants genetic response to moisture stress; two are ABA dependent while two are ABA independent. They all affect gene expression that increases the plants water stress tolerance.

The effects of moisture stress on photosynthesis can depend as much on the velocity and degree of photosynthetic recovery, as it depends on the degree and velocity of photosynthesis decline during water depletion. Plants that are subjected to mild stress can recover in 1–2 days however, plants subjected to severe water stress will only recover 40-60% of its maximum photosynthetic rates the day after re watering and may never reach maximum photosynthetic rates. The recovery from moisture stress starts with an increase in water content in leaves reopening the stomata then the synthesis of photosynthetic proteins.

## FLOOD STRESS

Flooding (also called as waterlogged/ponded/saturated/submerged soil) is one of the crucial stress factors that has an appreciable effect on crop growth, ultimately leading to reduced yield and production of different crop plants. It is usually induced as a result of climatic discrepancy, especially when there is an unpredictable, erratic, or irregular rainfall. This stress is mainly prevalent in the rain-fed ecosystem, where poor drainage occurs. Globally, nearly 13% of the total land and nearly 10% of the agricultural land are affected adversely because of the problem of flooding. The total yield loss in different crops suffering from waterlogging range from 15% to 80%, depending on the type of plant, kind of soil, and extent of the stress.

Different definitions of flooding have been given by researchers, but Perata et al. explained flooding as a type of complex stress factor, composed of ongoing cooperative changes within plant tissues triggered by accumulated water around the plant. Flooding is also defined as the complete saturation of soil with water, which facilitates anoxic conditions and provide good opportunity for anaerobes to perform anticipated activities more conveniently.

Most of the agriculturally essential crops are incompetent of tolerating this stress. Waterlogging severely affects the yield of farmland by restricting root development and thereby reducing the crop's ability to uptake nutrients and water. The Intergovernmental Panel on Climate Change stated that various anthropogenic activities have increased the frequency of tropical cyclone and heavy rain in plains and arable farmland affecting the poorest sections of the society. The impact of flooding is observed on a variety of crops, such as maize, oats, corn, wheat, rice, lentils, dry beans, soybean, lotus, peas, and peppers, and based on this, plants are either categorized as tolerant or sensitive toward flooding. Intolerant plants are excluded from flood-prone environment, whereas tolerant plants can be divided into plants that utilize an escape strategy and those that use a quiescence strategy. The escape strategy is based on a set of inducible anatomical and morphological characters that allow reaeration of flooded tissues, whereas the quiescence strategy is composed of characters that conserve the use of carbohydrates for prolonged underwater survival.

### Impact of Flooding

Flooded soil inflicts severe pressure on plant life, mainly because surplus water in the environment can deprive plants of their basic needs, such as $O_2$, $CO_2$, and sunlight. Reduced oxygen, sunlight, and carbon dioxide in turn hampers the process of photosynthesis and aerobic respiration, resulting in scarcity of carbohydrate thereby limiting growth and development of crop. Several researchers proved that excess of water results in complex changes such as impeded gas exchange, which subsequently affects the soil microbiome, leading to hypoxia and anoxia. Hypoxia encourages an increase in the redox potential between waterlogged soil and plants, which leads to the production of reactive oxygen species (ROS). ROS is necessary for intra/intercellular signaling, but as its concentration rises, it disrupts normal plant metabolism through oxidation of membrane lipids, pigments, proteins, and nucleic acids. A significant increase in the activity of superoxide radical (SOD) production, lipid peroxidation, and membrane injury was observed in mung, maize, and pigeon pea genotypes in response to flooding. An increase in the level of $CO_2$ in plant roots is accompanied by phytotoxin accumulation in reduced soil, which inhibits respiration. Furthermore, reoxygenation of organs and tissues once the flood subsides often leads to oxidative stress.

## Mechanisms Underlying Flood Tolerance

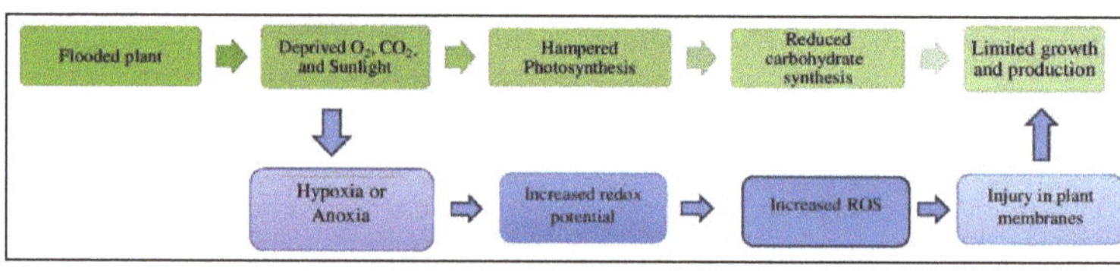


Delerious effects of flooding on plants.

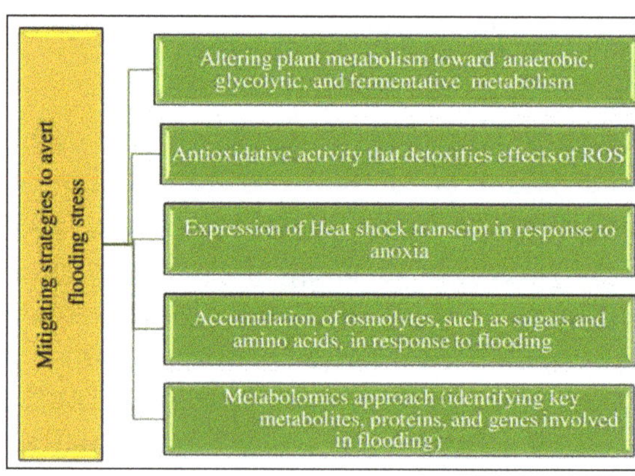


Mechanism to overcome the effects of flooding stress in plants.

Different plants display incredible diversity toward flooding, but an important question is that how these plants sense, acclimatize, and adapt toward flooding stress. Here we discuss a few mechanisms that help plants in mitigating the detrimental effect of flooding stress.

## Shift in Metabolism

During flooding, plants develop a set of morphological, physiological, and anatomical changes to to protect themselves from this detrimental stress. The most common morphological response toward flooding is the development of adventitious roots. These roots have a high porosity that helps in the uptake of water and nutrients under the submerged state. Expansion of adventitious roots helps in the oxygenation of submerged tissues through the aerenchyma tissue and thereby increases plant biomass. The generation of aerenchyma in tissues is an anatomical response, which enables the passage of oxygen from shoots to roots. At the physiological level, flooding alters plant carbon fixation and water retention. It also checks closing of stomata, the rate of transpiration, and monitors photosynthesis depending on the tolerance to flooding of each plant species. Expansion of adventitious roots in flooded tomato, rice, and Rumex is an ethylene-dependent process. Flooding-adapted Oryza has constitutively developed adventitious root primordia buried under nodal tissues. The interaction of auxin and ethylene leads to the emergence of adventitious roots in rice. Ethylene is an important regulator of several adaptive systemic responses of the waterlogging. 1-Aminocyclopropane-1-Carboxylate (ACC) is the precursor of ethylene; under the flooding state, ACC is transported from the roots to the shoots. This oxygen-mediated conversion of ACC to ethylene triggers adaptive changes typically observed in the shoots of waterlogged plants, such as shoot aerenchyma formation and leaf nastic movements. Plants such as Arabidopsis, rice, tomato, and Rumex exhibit leaf hyponasty during waterlogging to protect themselves from stress by an ethylene-dependent mechanism. Hyponastic growth increases anoxia tolerance by facilitating the leaves to rise above water level and coping with stress.

Plants also tolerate oxidative stress by shifting their metabolism or altering their anatomy. Ethylene responsive factors (ERFs), ERF VII, are a low oxygen sensing group that regulates adaptation toward flooding. Developmental adaptation comprises variation in organ and cellular composition for enhancing the accessibility of oxygen. During flooding when plants are unable to avail oxygen for aerobic respiration, they maintain cell viability by shifting toward anaerobic, glycolytic, and fermentative metabolism. The glycolytic pathway is less efficient and yields two ATP molecules instead of 36 ATP per hexose molecule formed during oxidative phosphorylation. Glycolysis is also limited by the rate of NADH oxidation. Ethanolic fermentation involving alcohol dehydrogenase (ADH) and lactic acid fermentation involving lactic dehydrogenase (LDH) are the major mechanisms of NADH oxidation working under anaerobic conditions. Energy-demanding progressions, such as replication, transcription, and cell division, are truncated during anoxic conditions, prevalent during flooding. The process of translation involving protein synthesis is finely regulated and encodes proteins involved in anaerobic and ROS metabolism. Different anaerobic proteins (ANPs), such as pyruvate decarboxylase, sucrose synthase, LDH, and ADH, which are synthesized in response to flooding stress.

## Antioxidative Activity

Submergence may cause plant senescence, because of the decrease in $CO_2$ concentration and increase in the concentration of ROS. The toxic ROS, including singlet oxygen ($1O_2$), superoxide $O_2$ –, hydroxyl radical (OH%), and hydrogen peroxide ($H_2O_2$), can oxidize proteins, lipids, carbohydrates, and nucleic acids. In addition, these radicals can inactivate various Calvin cycle enzymes that are involved in oxidative systems. To ameliorate the oxidative damage and protect tissues and cells, plants have developed a complex antioxidant defensive system. Enzymatic antioxidants, such as catalase, ascorbate peroxidase, SOD, dehydroascorbate reductase, mono-dehydroascorbate reductase, glutathione reductase, and glutathione S-transferase, and nonenzymatic antioxidants, including α-tocopherol, ascorbic acid, glutathione, and carotenoids, respectively, detoxify the detrimental effects of ROS. Upregulation of the antioxidant defense system is often correlated with flood tolerance. It was observed that exposing soybean plant to anoxic stress (of 2 h) increased the level of SODs; however, exposing the same plant to anoxia for a longer duration (of 5 h) increased its ability to cope with free oxygen radicals, and it exhibited less injury. Incubating plants with exogenous ascorbic acid alleviated postanoxic injury in soybean roots. Roots and leaves of Hordeum vulgare and wheat showed $H_2O_2$ accumulation under hypoxic conditions. The existence of $H_2O_2$ in the apoplast and plasma membrane has been visualized under hypoxic conditions by Transmission Electron Microscope (TEM) in different plant species. Indirect evidence of ROS formation (i.e., lipid peroxidation products) under low oxygen has also been reported by the same workers. An increase in the activity of NADPH oxidase and NADPH oxidase mRNA expression under hypoxic condition in pigeon pea genotypes was observed by Sairam et al. under flooding stress. Monk et al. reported a 14-fold increment in SOD activity during hypoxia in the Iris plant. Similarly, increased activity of glutathione reductase and glutathione content in wheat seedlings mitigated posthypoxia oxidative stress. Reduced activity of mono-dehydroascorbate reductase, dehydroascorbate reductase, and glutathione reductase enzymes was observed in wheat plants under conditions of anoxia. A study conducted by Thirunavukkarasu et al. showed transcriptomic changes in flood-tolerant and flood-susceptible maize genotypes by displaying an upregulation of flood-tolerant genes involved in antioxidant defense and fermentation pathway. Flood-tolerant pigeon pea genotypes showed an incessant increase in antioxidant activity after 6 days of flooding; however, in susceptible genotypes the same activity dropped within 2 days of waterlogging.

## Heat Shock Transcription Factors

Heat shock transcription factors (HSF) are also known as $H_2O_2$ sensors in plants. A rise in heat shock protein (HSP) in response to anoxia is conserved in various plant species. Banti et al. observed the transgenic model plant Arabidopsis expressing HSFA2 is more tolerant toward hypoxia compared with the wild type. Researchers also reported the induction of HSFs in response to flooding, which leads to the transcription of genes encoding both high and low-molecular-weight HSPs in plants. The NADPH oxidase–dependent ROS pathway regulates the ROS-related transcription factor and HSFs under the flooding state. Transcription factors regulated by HSPs and ROS are not N-end rule pathway–dependent proteolysis; hence ROS signaling is a self-regulating mechanism working under flood tolerance.

## Osmoprotectants

Flooding is an important stress factor that hampers the phenomena of photosynthesis. The concentration of amino acids, sugars, and sugar alcohols in plant tissue is affected by the impairment in $CO_2$ assimilation and by complex regulator. Plants accumulate different sugars and amino acids to protect themselves from stress. Osmoprotectants, such as glycine betaine and mannitol, play an important role in stress amelioration. Glycine betaine is copiously found in the chloroplast where it provides safety to the thylakoid membranes and preserves the photosynthetic efficiency. Mannitol acts as a free radical scavenger and stabilizes cellular and subcellular structures and plays a significant role in carbon storage during flooding stress. When rice plant was exposed to flooding, there was an increase in the proline content (under submerged conditions) in comparison with nonsubmerged conditions serving as control. The accumulation of proline under stress conditions (due to de novo synthesis) is an adaptive mechanism to maintain osmoregulation. Cabellero et al. reported the role of proline as a compatible solute that adjusts the osmotic potential in the cytoplasm under flooding state. A similar report of proline accumulation in response to flooding in black pepper was reported by Ou et al. Proline acts as an antidehydrating agent and a cytoplasmic penetrant and can reduce cell osmotic potential under flooded conditions. Proline acts as a compatible solute, and the multifunctional role of proline has been elucidated by several researchers in ROS scavenging, protein stabilization, and DNA stabilization. On the other hand, another group of workers, Wangkhem and Kananbala, stated a significant increase in the accumulation of proline and total sugars after exposing plant to complete submergence for 20 days. Gas chromatography (GC) coupled with electron-impact time-of-flight mass spectrometry (TOF-MS) can be used for the identification of these metabolites (sugars and amino acids). Submerged Arabidopsis plants harbor a high amount of soluble sugars and amino acids as a result of improved fatty acid and starch catabolism. The function of carbohydrate transporters was upregulated in shoots, following the transfer of carbohydrates from shoots to roots. Submergence lead to ROS production and the accumulation of sugars, such as sucrose and trehalose, that can ameliorate the effect of oxidative stress. During submergence, there is a decline in cytosolic pH and oxygen deprivation, which induces the greatest accumulation of gamma aminobutyric acid proteins. These proteins act as a stress mitigator and protect the plant from flood injuries. Carbohydrate metabolism plays a crucial role in flood tolerance, as it affects photosynthetic activity. Utilization of glucose, fructose, fructan, and starch as energy source was observed during the period of stress. These sugars also act as osmolytes and maintain cell turgidity and membrane stability, preventing the degradation of proteins. In addition, an increase in the concentration of sugar alcohols, such as sorbitol or mannitol, has also been linked to stress tolerance. Accumulation of sugars in the roots of Inga marginata was observed when seeds were exposed to flooding stress as reported by Bender et al. Qin et al. outlined that reduced consumption and enhancement in the accumulation of soluble sugars are the main mechanisms associated with flood tolerance.

## Role of Metabolomics in Flood Tolerance

As climatic conditions are turning hostile day by day, and the incidence of flooding is becoming more common, there is an emergent need for developing new resistant crop varieties for sustainability. This is possible only through genetic engineering and by understanding the detailed mechanisms involved in flooding stress. Combination of contemporary knowledge with traditional breeding approach is essential for improving existing plant varieties with improved technology to nonoptimal environmental conditions. The development of the "omics" technologies, such as genomics, proteomics, transcriptomics, and metabolomics, has transformed plant research and has allowed it holistically to integrate metabolites, proteins, genes, regulatory elements, and fluxes within it.

Metabolomics is a novel associate to the "omics" family of systems biology and is defined as a quantitative and comprehensive evaluation of all molecules in the biological system. In simple terms, metabolomics is a large-scale study of very small molecules (known as metabolites<1500Da) that are present within cells, tissues, and biofluids of organisms. All together, the cooperative functions of these small molecules and their interactions within the living system are known as the metabolomics. It is the widespread, nonbiased, high-throughput investigation of complex metabolite mixtures that allows preferably the identification and quantification of every single metabolite. Although metabolomics is new compared to proteomics and genomics, it is now firmly recognized as a purposeful genetic approach for understanding the molecular complexity of plant life. Metabolomic approaches provide a detailed insight into the metabolic pathways, their links, and connections with each other; it also clarifies the role of genes, proteins, and metabolites that participate in various aspects of functional genomics and systems biolog.

Metabolomics in an emerging technique that has drawn attention in different areas of plant science, including stress biology, phytochemistry, and environmental physiology. In general, as genomics is rooted in classical molecular biology so is metabolomics deeply related to phytochemistry. The objective of metabolomics study is qualitative and quantitative analysis of metabolites in a plant species at a given developmental stage under particular environmental conditions. Metabolomics study is used for several purposes such as checking the influence of abiotic stress factor on plant metabolism, tracking the effect of certain genes, proteins, and metabolites within the plants that are assisting in stress tolerance. The exact composition of metabolites in a given plant is the result of a particular gene expression profile. The activation of metabolic pathways in turn activates intermediates and precursors to produce a bioactive molecule, such as a signaling compound, an antioxidant molecule, a cell structure biosynthesis intermediate, and even a storage compound. The assembly of these complexes can be controlled in turn by other signal molecules that can positively or negatively regulate other metabolic steps. Under a stressed state, metabolites also damage the cell and malondialdehyde, lipid peroxides, and DNA fragments resulting from oxidative damage are produced. Considering the metabolome, the balance between the impaired metabolite, defense, and signaling can be used to evaluate plant tolerance in any stress conditions.

Several analytical techniques are involved in analyzing different plant metabolites under control or flooded conditions. A separation technique coupled with a detection device, usually MS, is used nowadays. However, there are other substitutes to separation techniques, such as flow injection analysis coupled to MS, or analyzers such as nuclear magnetic resonance (NMR) or Fourier transform infrared spectroscopy (FTIR) for fingerprinting purposes. Separation of metabolites involves selective chromatography, such as separation of volatile compounds and primary metabolites (sugars, amino acids, or tricarboxylic acid cycle intermediates) requires GC. Liquid chromatography (LC) is broadly used to analyze a mixed array of diverse secondary metabolites. Capillary electrophoresis also provides similar features as that of LC; in addition, it can also separate ionic metabolites. These techniques are used in combination since they are mostly complimentary to each independent technique having special coverage of assorted metabolite. Owing to the high diversity in chemical properties and structures of molecules, so far no sole method has been developed to classify, identify, and enumerate them. In spite of this the most comprehensive procedures distinguish only between 1000 and 2000 molecular features. However, more precise and targeted techniques for the quantitation of plant metabolites are needed to be explored for enhancing specificity and sensitivity in the results.

Several investigations have explained the functions of different key metabolites, genes, proteins, and molecular networks involved in flooding stress. Successive advancement in dense chromosome mapping allowed researchers to identify corresponding quantitative trait loci (QTLs), which has become a crucial molecular tool for plant breeding programs. In spite of all these new research, still there are lots of unknown molecular processes occurring in plants that assist in coping stress. Proteomic and metabolomic studies have now become commanding tools to screen the changes in response to numerous environmental stresses.

In Arabidopsis the expression of genes encoding ANPs were involved in sugar metabolism. In model crop Arabidopsis, genes encoding ANPs are somewhat involved in sugar metabolism. During hypoxic conditions, signal transduction components that activate calmodium, calmodium-associated proteins, mitochondrial alternative oxidase, and RopGAP4 were upregulated in Arabidopsis. The energy-sensing protein kinases KIN10 and KIN11 belonging to SnRK1 clade of Arabidopsis regulate carbon utilization under hypoxic condition. KIN10 positively regulates genes encoding enzymes that catalyze carbohydrate and amino acid catabolism in Arabidopsis. One of the KIN10 or KIN11-regulated genes is EXORDIUM-LIKE1, an Hypoxiaresponsive unknown protein (HUP) that is essential for carbon management under low-oxygen. Under low-oxygen stress, ROS have connection with the signal transduction pathway. Screening of gene trap transposon (Ds-Gus) from Arabidopsis led to the identification of mutants with increased ADH activity, under hypoxic conditions. The mutant phenotype is formed from the insert of DsGus in the first exon of a gene that encodes Rop (RHO-like small G protein of plants) guanosine triphosphatase (GTPase) activating protein, ROPGAP4. Rop signaling is related to $H_2O_2$ and $Ca^{2+}$ signaling for flood tolerance. Stimulation of ROP by GTP under low $O_2$ induces the production of $H_2O_2$ through a calcium-dependent NADPH oxidase. Accumulation of $H_2O_2$ is crucial for the induction of ADH and RopGAP4 expression. The level of Rop is regulated by RopGAP4, which participates in negative feedback regulation ROP activity; hence, RopGAP4-1 seedlings fail to resist oxidative stress due to $H_2O_2$ accumulation.

Wang et al. elucidated the impact of flooding on soybean seedling by performing organ-specific analysis through a gel-free/label-free proteomic technique. The activities of ADH and delta-1-pyrroline-5-carboxylate synthase were noticeably increased in roots and leaves of plants exposed to flooding. The tricarboxylic acid cycle was completely suppressed in leaves and roots, whereas proteins related to DNA, RNA signaling, and photosynthesis were badly affected in leaf and roots in response to flooding. Seventeen different proteins, including β-glucosidase 31 and β-amylase 5, were observed in leaves and roots of soybean seeds during flooding. The expression of β-amylase 5 might be involved in carbohydrate mobilization for providing energy to the leaf of soybean plant.

In the paddy field, flooding is regulated by QTL found on chromosome 9. The protein SUB1A is an ethylene-responsive transcription factor that limits the growth of crop under flooding state. Metabolic profiling of the crossbred line M202 (Sub1) displayed high tolerance toward flood in comparison with wild strain M202. This suggests that the presence of SUB1A in M202 led to the repression of carbohydrate metabolism in shoot tissues. These findings recommend that in the crossbred line M202 (Sub1) with SUB1A, carbohydrate metabolism is arranged in such a manner that it suppresses growth if the plant is submerged and further reduces energy loss under unfavorable conditions. Neeraja et al. also used molecular markers for Sub1 gene in a backcross breeding program with recurrent parent Swarna. This Sub1 provides tolerance to the flood-sensitive mega varieties. This QTL has provided a great opportunity for marker-assisted backcrossing for developing submergence tolerance in mega varieties. Singh et al. identified unique QTLs and discuss their physiological consequences.

The complete transcriptome of maize at three different stages of flooding was analyzed by Thirunavukkarasu et al. Genes responsible for programmed cell death following aerenchyma formation was selectively upregulated in the HKI 1105 tolerant genotype after submergence. Calcium-binding proteins, such as calmodulin, were overexpressed in HKI 1105 and interact with glutamate decarboxylase, which in turn helps in maintaining the cytosolic pH under anoxia. The gene specific for flooding, XET A, was upregulated in HKI 1105 during severe (16-fold) and moderate (253-fold) stresses, but downregulated in V372 (sensitive). Genes involved in signal transduction and plant hormone biosynthesis were differentially controlled under waterlogged state. There was an increase in the level of ethylene, indole-3-acetic acid (IAA), abscisic acid, and gibberellic acid and a reduction in the production of cytokinin. ERFs, such as BBM2, protein 1, AIL5-like, and WRI1, were upregulated in HKI 1105. In the tolerant genotype, auxin receptor genes, such as IAA14, IAA16, and IAA3, were unregulated. Signaling pathways between auxin and ethylene increase the formation of lateral and adventitious roots in the flood-tolerant genotype of maize.

Rumex palustris shows hyponastic growth by orientating its leaves in an ethylene-dependent manner. In Rumex acetosa, accumulation of ethylene reduces petiole elongation and inclines the plants toward metabolic rearrangement to minimize carbon consumption. Van Veen et al. employed RNA sequencing technology to investigate the molecular basis of adaptive traits of these two species. There was an enhanced expression of ACC oxidase, enzyme catalyzing ethylene biosynthesis in both the species. In R. palustris, the expression of an EIN3 BINDING F-BOX was specifically upregulated. Complete metabolic reprogramming of R. acetosa enhanced its survivability in flood-prone area.

Luffa (Luffa cylindrica) is a flood-tolerant plant species. Su and Su monitored the metabolic responses of luffa roots toward flooding stress. It was observed that the respiratory activity of flooded roots reduced to 60% in comparison with the nonflooded ones. The activity of pyruvate decarboxylase and ADH amplified quickly. After exposing the plant to 7days of flooding, an approximately 50-fold increment in the level of both the enzymes was observed, which indicated strong alcoholic fermentation. Yet, the level of ethanol did not increase considerably, but declined after 1day. A high ADH/pyruvate decarboxylase ratio might play an important role in avoiding the accumulation of toxic acetaldehyde. In other fermentative pathways, the activity of phosphoenolpyruvate carboxylase, the key enzyme catalyzing malate synthesis, showed no measurable difference between nonflooded and flooded roots. NADP malic enzyme, catalyzing the decarboxylation of malate, displayed an increase in activity in flooded roots as compared with the nonflooded ones; the amount of malate in flooded roots reduced to 40% of the control ones.

Water lily Nymphoides peltata is widely distributed in subtropical and temperate regions of Eurasia. Leaves of this plant float on the water surface with connecting petioles and roots are fixed to the bottom mud. As this plant is a nonmodel organism genomic information is very less; so, global gene expression changes of N. peltata was monitored both under control and flooded conditions using Illumina RNA-Seq technology. The results gave a comprehensive view of the complex molecular events involved in the response of floating-leaved plants toward flooding. The results of Illumina RNA-Seq technology gave 78,037,588 and 103,266,542 clean reads from the untreated sample (US) and the treated sample (TS), respectively. Association of the reads from TS and US individually generated 95,372 and 87,673 unigenes, respectively. Pooling of all clean reads from US and TS together constituted 124,302 unigenes with an N50 length of 1449bp after assembly. The number of unigenes from the pooled strategy was observed to be larger than those from separate assemblies, because some unigenes with low expression levels generated with the mutual approach cannot be found when using the strategy of single approach because of fewer reads. In total, 8883 unigenes were differentially expressed between the normal and flooded states. Among them, most differentially expressed genes were primarily related to photosynthesis, antioxidant process, and carbohydrate metabolism (including glycolysis, tricarboxylic acid cycle, and alanine metabolism). The antioxidative activity and glycolysis are usually induced by a flooded environment, whereas alanine metabolism and photosynthesis are rarely found in flooding-tolerant plants. These results suggest the significant contributions of both the processes in the active response of N. peltata toward flooding stress.

# DROUGHT STRESS

Faced with scarcity of water resources, drought is the single most critical threat to world food security. It was the catalyst of the great famines of the past. Because the world's water supply is limiting, future food demand for rapidly increasing population pressures is likely to further aggravate the effects of drought. The severity of drought is unpredictable as it depends on many factors such as occurrence and distribution of rainfall, evaporative demands and moisture storing capacity of soils.

Investigations carried out in the past provide considerable insights into the mechanism of drought tolerance in plants at molecular level. Three main mechanisms reduce crop yield by soil water deficit:

- Reduced canopy absorption of photosynthetically active radiation.
- Decreased radiation-use effciency.
- Reduced harvest index.

The reproducibility of drought stress treatments is very cumbersome, which significantly impedes research on plant drought tolerance. A slow pace in revealing drought tolerance mechanisms has hampered both traditional breeding efforts and use of modern genetics approaches in the improvement of drought tolerance of crop plants. Although plant responses to drought are relatively well known, plant performance under a more complex environment where multiple stresses co-occur is fragmentary. That is why the plants have to respond simultaneously to multiple stresses, e.g. drought, excessive light and heat, which may coincide in the field. These kinds of investigations are usually not predictable from single factor studies.

It is imperative to improve the drought tolerance of crops under the changing circumstances. Currently, there are no economically viable technological means to facilitate crop production under drought. However, development of crop plants tolerant to drought stress might be a promising approach, which helps in meeting the food demands. Development of crops for enhanced drought resistance, among other things, requires the knowledge of physiological mechanisms and genetic control of the contributing traits at different plant developmental stages. Valuable work has been done on drought tolerance in plants. Ingram and Bartels more than a decade ago elegantly reviewed those appreciable efforts. More recent reviews deal with specific aspects of plant drought tolerance.

## Effects of Drought in Plants

The effects of drought range from morphological to molecular levels and are evident at all phenological stages of plant growth at whatever stage the water deficit takes place.

An account of various drought stress effects and their extent is elaborated below.

## Crop Growth and Yield

The first and foremost effect of drought is impaired germination and poor stand establishment. Drought stress has been reported to severely reduce germination and seedling stand. In a study on pea, drought stress impaired the germination and early seedling growth of five cultivars tested. Moreover, in alfalfa (Medicago sativa), germination potential, hypocotyl length, and shoot and root fresh and dry weights were reduced by polyethylene glycol-induced water deficit, while the root length was increased. However, in rice, drought stress during the vegetative stage greatly reduced the plant growth and development.

Growth is accomplished through cell division, cell enlargement and differentiation, and involves genetic, physiological, ecological and morphological events and their complex interactions. The quality and quantity of plant growth depend on these events, which are affected by water deficit. Cell growth is one of the most drought-sensitive physiological processes due to the reduction in turgor pressure. Under severe water deficiency, cell elongation of higher plants can be inhibited by interruption of water flow from the xylem to the surrounding elongating cells.

Effect of drought stress on the vegetative growth of rice cv. IR64. Both the plants were grown under well-watered conditions up to 20 days following emergence. One pot was submitted to progressive soil drying (drought stress). The afternoon before the drought, all pots were fully watered (to saturation). After draining overnight, the pots were enclosed around the stem to prevent direct soil evaporation. A small tube was inserted for re-watering pots. The decrease in soil moisture was controlled by partial re-watering of the stressed pots to avoid a quicker imposition of stress and to homogenize the development of drought stress. A well-watered control pot was maintained at the initial target weight by adding the daily water loss back to the pot. This figure shows the plants 20 days after imposition of drought stress.

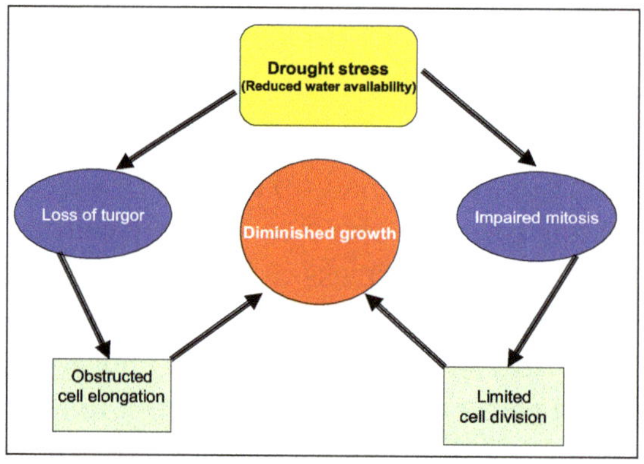


Description of possible mechanisms of growth reduction under drought stress. Under drought stress conditions, cell elongation in higher plants is inhibited by reduced turgor pressure. Reduced water uptake results in a decrease in tissue water contents. As a result, turgor is lost. Likewise, drought stress also trims down the photoassimilation and metabolites required for cell division. As a consequence, impaired mitosis, cell elongation and expansion result in reduced growth.

Table: Impaired mitosis, cell elongation and expansion result in reduced plant height, leaf area and crop growth under drought.

| Crop | Growth stage | Yield reduction |
|---|---|---|
| Barley | Seed filling | 49–57% |
| Maize | Grain filling | 79–81% |
| Maize | Reproductive | 63–87% |
| Maize | Reproductive | 70–47% |
| Maize | Vegetative | 25–60% |
| Maize | Reproductive | 32–92% |
| Rice | Reproductive (mild stress) | 53–92% |
| Rice | Reproductive (severe stress) | 48–94% |
| Rice | Grain filling (mild stress) | 30–55% |
| Rice | Grain filling (severe stress) | 60% |
| Rice | Reproductive | 24–84% |
| Chickpea | Reproductive | 45–69% |
| Pigeonpea | Reproductive | 40–55% |
| Common beans | Reproductive | 58–87% |
| Soybean | Reproductive | 46–71% |
| Cowpea | Reproductive | 60–11% |
| Sunflower | Reproductive | 60% |
| Canola | Reproductive | 30% |
| Potato | Flowering | 13% |

Many yield-determining physiological processes in plants respond to water stress. Yield integrates many of these physiological processes in a complex way. Thus, it is difficult to interpret how plants accumulate, combine and display the ever-changing and indefinite physiological processes over the entire life cycle of crops. For water stress, severity, duration and timing of stress, as well as responses of plants after stress removal, and interaction between stress and other factors are extremely important. For instance, water stress applied at pre-anthesis reduced time to anthesis, while at post-anthesis it shortened the grain-filling period in triticale geno-types. In barley (Hordeum vulgare), drought stress reduced grain yield by reducing the number of tillers, spikes and grains per plant and individual grain weight. Post-anthesis drought stress was detrimental to grain yield regardless of the stress severity.)

Drought-induced yield reduction has been reported in many crop species, which depends upon the severity and duration of the stress period. In maize, water stress reduced yield by delaying silking, thus increasing the anthesis-to-silking interval. This trait was highly correlated with grain yield, specifically ear and kernel number per plant. Following heading, drought had little effect on the rate of kernel filling in wheat, but its duration (time from fertilization to maturity) was shortened and dry weight reduced at maturity. Drought stress in soybean reduced total seed yield and the branch seed yield. In pearl millet (Pennisetum glaucum), co-mapping of the harvest index and panicle harvest index with grain yield revealed that greater drought tolerance was achieved by greater partitioning of dry matter from stover to grains.

Drought at flowering commonly results in barrenness. A major cause of this, though not the only one, was a reduction in assimilate flux to the developing ear below some threshold level necessary to sustain optimal grain growth. Moisture deficit reduced cotton (Gossypium hirsutum) lint yield, although the timing, duration, severity and speed of development undoubtedly had pivotal roles in determining how the plant responded to moisture deficit. Lint yield was generally reduced due to reduced boll production because of fewer flowers and greater boll abortions when the stress intensity was greater during reproductive growth.

Grain filling in cereals is a process of starch biosynthesis from simple carbohydrates. It is believed that four enzymes play key roles in this process: sucrose synthase, adenosine diphosphate-glucose-pyrophosphorylase, starch synthase and starch branching enzyme. Decline in the rate of grain growth resulted from reduced sucrose synthase activity, while cessation of growth resulted from inactivation of adenosine diphosphate-glucose-pyrophosphorylase in the water-stressed wheat. Water deficit during pollination increased the frequency of kernel abortion in maize (Zea mays). Under water stress, diminished grain set and kernel growth in wheat and a decreased rate of endosperm cell division was associated with elevated levels of abscisic acid in maize. In pigeonpea, drought stress coinciding with the flowering stage reduced seed yield by 40–55%. In rice, on the other hand, water stress imposed during the grain-filling period enhanced remobilization of pre-stored carbon reserves to grains and accelerated grain filling. In summary, prevailing drought reduces plant growth and development, leading to hampered flower production and grain filling, and thus smaller and fewer grains. A reduction in grain filling occurs due to a reduction in the assimilate partitioning and activities of sucrose and starch synthesis enzymes.

## Water Relations

Relative water content, leaf water potential, stomatal resistance, rate of transpiration, leaf temperature and canopy temperature are important characteristics that influence plant water relations. Relative water content of wheat leaves was higher initially during leaf development and decreased as the dry matter accumulated and leaf matured. Obviously, water-stressed wheat and rice plants had lower relative water content than non-stressed ones. Exposure of these plants to drought stress substantially decreased the leaf water potential, relative water content and transpiration rate, with a concomitant increase in leaf temperature. A conservative influence of decreased stomatal conductance in non-irrigated plants was negated by a leaf-to-air vapor pressure difference caused by the associated higher leaf temperature. Transpiration rates were similar in both treatments and the lower total water use of the non-irrigated stand resulted entirely from a smaller leaf area index.

Nerd and Nobel reported that during drought stress, total water contents of Opuntia ficus-indica cladode were decreased by 57%. The water-storage parenchyma of the cladodes lost a greater fraction of water than the chlorenchyma, and thus showed a lower turgor potential. In another study on Hibiscus rosa-sinensis, relative water content, turgor potential, transpiration, stomatal conductance and water-use efficiency were decreased under drought stress.

The ratio between dry matter produced and water consumed is termed as water-use efficiency at the whole-plant level. Abbate et al. concluded that under limited supply, water-use efficiency of wheat was greater than in well-watered conditions. They correlated this higher water-use efficiency with stomatal closure to reduce the transpiration. In another study on clover (Trifolium alexan-drinum), water-use efficiency was increased due to lowered water loss under drought stress, primarily by decreased transpiration rate and leaf area, and relatively lesser reduction in yield. Also, in Pinus ponderosa and Artemisia tridentata, drought stress did not reduce the water-use efficiency; rather, it was increased, mainly due to a rapid decrease in stomatal conductance with increasing water deficit. Lazaridou et al. further reported that leucern (Medicago sativa) grown under drought had greater water-use efficiency than that under irrigated conditions, for the same leaf water potential. However, in potato, early season drought stress significantly minimized the water-use efficiency, leading to greatly decreased growth and biomass accumulation.

In fact, although components of plant water relations are affected by reduced availability of water, stomatal opening and closing is more strongly affected. Moreover, change in leaf temperature may be an important factor in controlling leaf water status under drought stress. Drought-tolerant species maintain water-use efficiency by reducing the water loss. However, in the events where plant growth was hindered to a greater extent, water-use efficiency was also reduced significantly

## Nutrient Relations

Decreasing water availability under drought generally results in limited total nutrient uptake and their diminished tissue concentrations in crop plants. An important effect of water deficit is on the acquisition of nutrients by the root and their transport to shoots. Lowered absorption of the inorganic nutrients can result from interference in nutrient uptake and the unloading mechanism, and reduced transpirational flow. However, plant species and genotypes of a species may vary in their response to mineral uptake under water stress. In general, moisture stress induces an increase in N, a definitive decline in P and no definitive effects on K.

Transpiration is inhibited by drought, as shown for beech, but this may not necessarily affect nutrient uptake in a similar manner. Influence of drought on plant nutrition may also be related to limited availability of energy for assimilation of $NO^{-}3$ /$NH^{+}$, $PO^{3-}$ and $SO^{2-}$: they must be converted in energy-dependent processes before these ions can be used for growth and development of plants.

As nutrient and water requirements are closely related, fertilizer application is likely to increase the efficiency of crops in utilizing available water. This indicates a significant interaction between soil moisture deficits and nutrient acquisition. Studies show a positive response of crops to improved soil fertility under arid and semi-arid conditions. Currently, it is evident that crop yields can be substantially improved by enhancing the plant nutrient efficiency under limited moisture supply. It was shown that N and K uptake was hampered under drought stress in cotton.

Likewise, P and $PO^{3-}$ contents in the plant tissues diminished under drought, possibly because of lowered $PO^{3-}$ mobility as a result of low moisture availability. In drought-treated sunflower, the degree of stomatal opening of $K^+$-applied plants initially indicated quicker decline. However, at equally low soil water potential, diffusive resistance in the leaves of $K^+$-applied plants remained lower than those receiving no $K^+$. In summary, drought stress reduces the availability, uptake, translocation and metabolism of nutrients. A reduced transpiration rate due to water deficit reduces the nutrient absorption and efficiency of their utilization

## Photosynthesis

A major effect of drought is reduction in photosynthesis, which arises by a decrease in leaf expansion, impaired photosynthetic machinery, premature leaf senescence and associated reduction in food production. When stomatal and non-stomatal limitations to photosynthesis are compared, the former can be quite small. This implies that other processes besides $CO_2$ uptake are being damaged. The role of drought-induced stomatal closure, which limits $CO_2$ uptake by leaves, is very important. In such events, restricted $CO_2$ availability could possibly lead to increased susceptibility to photo-damage.

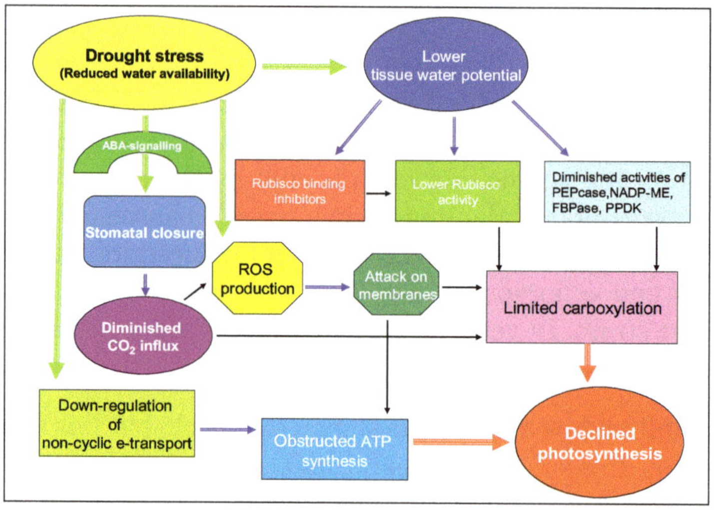


Photosynthesis under drought stress. Possible mechanisms in which photosynthesis is reduced under stress. Drought stress disturbs the balance between the production of reactive oxygen species and the antioxidant defense, causing accumulation of reactive oxygen species, which induces oxidative stress. Upon reduction in the amount of available water, plants close their stomata (plausibly via ABA signaling), which decreases the $CO_2$ influx. Reduction in $CO_2$ not only reduces the carboxylation directly but also directs more electrons to form reactive oxygen species. Severe drought conditions limit photosynthesis due to a decrease in the activities of ribulose-1, 5-bisphosphate carboxylase/oxygenase (Rubisco), phosphoenolpyruvate carboxylase (PEPCase), NADP-malic enzyme (NADP-ME), fructose-1, 6-bisphosphatase (FBPase) and pyruvate orthophosphate dikinase (PPDK). Reduced tissue water contents also increase the activity of Rubisco binding inhibitors. Moreover, non-cyclic electron transport is down-regulated to match the reduced requirements of NADPH production and thus reduces the ATP synthesis. ROS: reactive oxygen species.

Drought stress produced changes in photosynthetic pigments and components, damaged photosynthetic apparatus and diminished activities of Calvin cycle enzymes, which are important causes of reduced crop yield. Another important effect that inhibits the growth and photosynthetic abilities of plants is the loss of balance between the production of reactive oxygen species and the antioxidant defense, causing accumulation of reactive oxygen species which induces oxidative stress in proteins, membrane lipids and other cellular components. Some important components of photosynthesis affected by drought are discussed below.

## Stomatal Oscillations

The first response of virtually all plants to acute water deficit is the closure of their stomata to prevent the transpirational water loss. This may result in response to either a decrease in leaf turgor and water potential or to a low-humidity atmosphere. The debate as to whether drought mainly limits photosynthesis through stomatal closure or metabolic impairment has continued for a long time. During the last decade, stomatal closure was generally accepted to be the main determinant for decreased photosynthesis under mild to moderate drought.

When the amount of available soil water is moderately or severely limiting, the first option for plants is to close stomata. This decreases the inflow of $CO_2$ into the leaves and spares more electrons for the formation of into the leaves and spares more electrons for the formation of active oxygen species. As the rate of transpiration decreases, the amount of heat that can be dissipated increases. Various experiments have shown that stomatal responses are often more closely linked to soil moisture content than to leaf water status. This suggested that stomata respond to chemical signals, e.g. abcissic acid, produced by dehydrating roots, whilst leaf water status is kept constant. Environmental conditions that enhance the rate of transpiration also increase the pH of leaf sap, which can promote abscisic acid accumulation and concomitantly diminish stomatal conductance. Increased cytokinin concentration in the xylem sap promotes stomatal opening directly and affects the sensitivity of stomata towards abscisic acid.

Comparing results from different studies is complex due to interspecific differences in the response of stomatal conductance and photosynthesis to leaf water potential and relative water content; the parameters most often used to assess the degree of drought. It is clear that stomata close progressively as drought progresses, followed by a parallel decline in net photosynthesis. However, stomatal conductance is not controlled by soil water availability alone, but by a complex interaction of intrinsic and extrinsic factors.

## Photosynthetic Enzymes

Very severe drought conditions limit photosynthesis due to a decline in Rubisco activity. The activity of the photosynthetic electron transport chain is finely tuned to the availability of $CO_2$ in the chloroplast and change in photo-system II under drought conditions. Dehydration results in cell shrinkage, and consequently a decline in cellular volume. This makes cellular contents more viscous. Therefore, an increase in the probability of protein-protein interaction leads to their aggregation and denaturation. Increased concentration of solutes, leading to increased viscosity of the cytoplasm, may become toxic and may be deleterious to the functioning of enzymes, including those of the photosynthetic machinery.

The level of Rubisco in leaves is controlled by the rate of synthesis and degradation. Even under drought stress the Rubisco holoenzyme is relatively stable with a half-life of several days. However, drought stress showed a rapid diminution in the abundance of Rubisco small subunit transcripts, which indicated its decreased synthesis. Rubisco activity is modulated in vivo either by reaction with $CO_2$ and $Mg^{2+}$ to carbamylate a lysine residue in the catalytic site, or by binding inhibitors within the catalytic site. Such a binding either blocks activity or the carbamylation of the lysine residue, which is essential for activity. At night, 2-carboxyarabinitol-1-phosphate is formed in many species, which binds tightly to Rubisco, inhibiting catalytic activity. It is reported that tight-binding inhibitors can decrease Rubisco activity in the light. In tobacco (Nicotiana tabacum), decrease in Rubisco activity under drought stress was not a primary result of changes in activation by $CO_2$ and $Mg^{2+}$, and was rather due to the presence of tight-binding inhibitors. A rapid decline in photosynthesis under drought was accompanied by decreased maximum velocity of ribulose-1, 5-bisphosphate carboxylation by Rubisco, speed of ribulose-1, 5-bisphosphate regeneration, Rubisco and stromal fructose bis-phosphatase activities, and the quantum efficiency of photosystem II in higher plants. Moreover, under severe drought, carboxylation efficiency by Rubisco was greatly declined, and it acted more as oxygenase than carboxylase.

During water stress, activities of the phosphoenolpyruvate carboxylase, nicotinamide adenine dinucleotide phosphate-malic enzyme, Rubisco, fructose-1, 6-bisphosphatase and pyruvate orthophosphate dikinase decreased linearly with lowered leaf water potential. Pyruvate orthophosphate dikinase activities were decreased 9.1 times during water stress; a much greater reduction than other enzymes, which were from 2 to 4 times, suggesting that pyruvate orthophosphate dikinase is very likely to be the limiting enzyme to photosynthesis under water stress.

## Adenosine Triphosphate Synthesis

There is a long-standing controversy as to whether drought mainly limits photosynthesis through stomatal closure or by metabolic impairment. Evidence that impaired adenosine triphosphate synthesis is the main factor limiting photosynthesis even under mild drought has further stimulated the debate. It is reported that impaired photophosphorylation and adenosine triphosphate synthesis are the main factors limiting photosynthesis even under mild drought.

Under drought stress, production of limited nicotinamide adenine dinucleotide phosphate maintains the continuation of electron transport, although the status of the reductant may be high even when the fluxes are small, leading to a more increased demand than supply. Under drought stress, non-cyclic electron transport is down-regulated to match the requirements of decreased nicotinamide adenine dinucleotide phosphate production and cyclic electron transport is activated. This generates a proton gradient that induces the protective process of high-energy-state quenching. Support for this model came from the isolation of a mutant deficient in high-energy-state quenching that lacked cyclic electron transport. Support for cyclic electron transport under drought also came from non-steady-state measurements.

Dissipation mechanisms of excess photon energy under water stress were studied in ndhB-inactivated tobacco (cv. Xanthi) mutants impaired in reduced nicotinamide adenine dinucleotide phosphate dehydrogenase-dependent cyclic electron flow around photosystem I. The relative water content and net $CO_2$ assimilation was reduced to 30% and almost zero after an 11-day water stress regime in the mutant and wild-type plants, respectively. A decline in photosystem II activity (75%), and an increase in malondialdehyde (45%), an estimate of lipid peroxidation, were found in both the plant groups when subjected to water stress. Thus, a deficiency in reduced nicotinamide adenine dinucleotide phosphate dehydrogenase-dependent cyclic electron flow around photosystem I did not lead to oxidative damage because the mutant compensated for this deficiency by activating alternative dissipating routes of excess photon energy such as up-regulation of ferredoxin-dependentcyclic electron flow around photosystem I and enhanced accumulation of α-tocopherol (α-toc) quinine.

In fact, the activities of the enzymes of carbon assimilation and those involved in adenosine triphosphate synthesis are retarded and sometimes inhibited depending upon the extent of available moisture. Of these, Rubisco, which shows dual functions, acts as oxygenase under water-limiting conditions; and therefore limited $CO_2$ fixation is noticed.

## Assimilate Partitioning

Assimilate translocation to reproductive sinks is vital for seed development. Seed set and filling can be limited by availability or utilization, i.e., assimilate source or sink limitation, respectively. Drought stress frequently enhances allocation of dry matter to the roots, which can enhance water uptake. De Souza and Da Silv, while analyzing the partitioning and distribution of photo-assimilates in annual and perennial cotton under drought stress, reported that the root-to-shoot dry matter ratio was high in perennial cotton, thereby showing a preferential accumulation of starch and dry matter in roots as an adaptation to drought. Thus, perennial cotton apparently owed its drought resistance to the partitioning of assimilates that favored starch accumulation and growth of the root system. The export rate of sucrose from source to sink organs depends upon the current photosynthetic rate and the concentration of sucrose in the leaves. Drought stress decreases the photosynthetic rate, and disrupts the carbohydrate metabolism and level of sucrose in leaves that spills over to a decreased export rate. This is presumably due to drought stress-induced increased activity of acid invertase. Limited photosynthesis and sucrose accumulation in the leaves may hamper the rate of sucrose export to the sink organs and ultimately affect the reproductive development.

Apart from source limitation, the capacity of the reproductive sinks to utilize the incoming assimilates is also affected under drought stress and may also play a role in regulating reproductive abortion. Drought-induced carbohydrate deprivation, enhanced endogenous abscisic acid concentration, and an impaired ability to utilize the incoming sucrose by the reproductive sinks are potential factors contributing to seed abortion in grain crops. A reduced acid invertase activity can arrest the development of reproductive tissues due to improper phloem unloading. In addition, drought stress may inhibit important functions of vacuolar invertasemediated sucrose hydrolysis and osmotic potential modulation. In drought-stressed maize, a low invertase activity in the young ovaries lowers the ratio of hexoses to sucrose. This may inhibit cell division in the developing embryo/endosperm, resulting in weak sink intensity, and may ultimately lead to fruit abortion.

In summary, drought stress not only limits the size of the source and sink tissues but the phloem loading, assimilate translocation and dry matter portioning are also impaired. However, the extent of effects varies with the plant species, stage, duration and severity of drought.

## Respiration

Drought tolerance is a cost-intensive phenomenon, as a considerable quantity of energy is spent to cope with it. The fraction of carbohydrate that is lost through respiration determines the overall metabolic efficiency of the plant. The root is a major consumer of carbon fixed in photosynthesis and uses it for growth and maintenance, as well as dry matter production. Plant growth and developmental processes as well as environmental conditions affect the size of this fraction (i.e. utilized in respiration). However, the rate of photosynthesis often limits plant growth when soil water availability is reduced. A negative carbon balance can occur as a result of diminished photosynthetic capacity during drought, unless simultaneous and proportionate reductions in growth and carbon consumption take place.

In wheat, depending on the growth stage, cultivar and nutritional status, more than 50% of the daily accumulated photosynthates were transported to the root, and around 60% of this fraction was respired. Droughtsensitive spring wheat used a relatively greater amount of glucose to absorb water, especially in severe drought stress. Severe drought reduced the shoot and root biomass, photosynthesis and root respiration rate. Limited root respiration and root biomass under severe soil drying can improve growth and physiological activity of drought-tolerant wheat, which is advantageous over a drought-sensitive cultivar in arid regions.

There are two mitochondrial electron transport pathways from ubiquinone to oxygen in plants. The alternative pathway branches from the cytochrome pathway and donates electrons to oxygen directly by alternative oxidase.

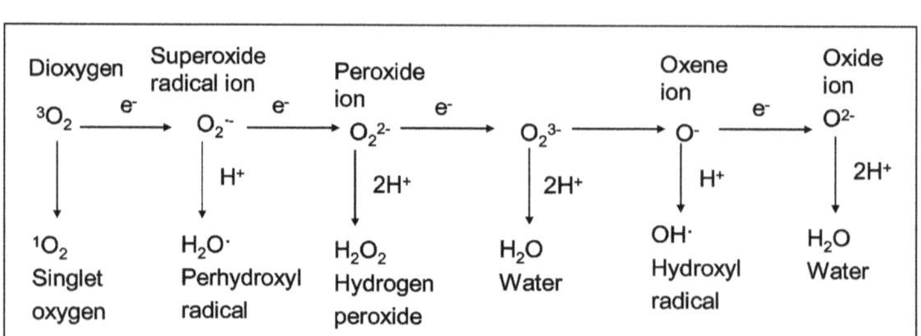


Generation of reactive oxygen species by energy transfer or sequential univalent reduction of ground state triplet oxygen.

The alternative pathway is not coupled with adenosine triphosphate synthesis, but can be induced in response to stress or inhibition of the main electron transfer pathway. When plants are exposed to drought stress, they produce reactive oxygen species, which damage membrane components. In this regard, alternative oxidase activity could be useful in maintaining normal levels of metabolites and reduce reactive oxygen species production during stress. Oxygen uptake by sugar beet was characterized by a high rate, distinct cytochrome oxidasedependent terminal oxidation and up to 80% inhibition of respiration in the presence of 0.5 mM potassium cyanide. At an early drought stage (10 days), a decrease in the activity of the cytochrome-mediated oxidation pathway was largely counterbalanced by the activation of mitochondrial alternative oxidase, whereas long-term dehydration of plants was accompanied by activation of additional oxidative systems insensitive to both potassium cyanide and salicylhydroxamate. In summary, water deficit in the rhizosphere leads to an increased rate of root respiration, leading to an imbalance in the utilization of carbon resources, reduced production of adenosine triphosphate and enhanced generation of reactive oxygen species.

## Oxidative Damage

Exposure of plants to certain environmental stresses quite often leads to the generation of reactive oxygen species, including superoxide anion radicals ($O^{-2}$), hydroxyl radicals (OH), hydrogen peroxide ($H_2O_2$), alkoxy radicals (RO) and singlet oxygen ($O^1$). Reactive oxygen species may react with proteins, lipids and de-oxyribonucleic acid, causing oxidative damage and impairing the normal functions of cells. Many cell compartments produce reactive oxygen species; of these, chloroplasts are a potentially important source because excited pigments in thylakoid membranes may interact with $O_2$ to form strong oxidants such as $O^{-2}$ or $O^1$. Further downstream reactions produce other reactive oxygen species such as $H_2O_2$ and $OH^-$. The interaction of $O_2$ with reduced components of the electron transport chain in mitochondria can lead to reactive oxygen species formation, and peroxisomes produce $H_2O_2$ when glycolate is oxidized into glyoxylic acid during photorespiration.

Mechanisms for the generation of reactive oxygen species in biological systems are represented by both non-enzymatic and enzymatic reactions. The partition between these two pathways under oxygen deprivation stress can be regulated by the oxygen concentration in the system. In non-enzymatic reactions, electron $O_2$ reduction can occur at higher oxygen concentrations. At very low $O_2$ concentrations, plant terminal oxidases and the formation of reactive oxygen species via the mitochondrial electron transport chain still remain functional. Among enzymatic sources of reactive oxygen species, xanthine oxidase, an enzyme responsible for the initial activation of $O_2$, should be mentioned. The electron donor xanthine oxidase can use xanthine, hypoxanthine or acetaldehyde, while the latter has been shown to accumulate under oxygen deprivation. This can represent a possible source for hypoxia-stimulated reactive oxygen species production. The next enzymatic step is the dismutation of the superoxide anion by superoxide dismutase to yield $H_2O_2$. Peroxidases and catalases also play an important role in the fine regulation of reactive oxygen species in the cell through activation and deactivation of $H_2O_2$. Several apoplastic enzymes may also generate reactive oxygen species under normal and stressful conditions. Other oxidases, responsible for the two-electron transfer to dioxygen (amino acid oxidases and glucose oxidase) can contribute to $H_2O_2$ accumulation.

Reactive oxygen species are formed as by-products in the electron transport chains of chloroplasts, mitochondria and the plasma membrane. The plant mitochondrial electron transport chain, with its redox-active electron carriers, is considered as the most probable candidate for intracellular reactive oxygen species formation. Mitochondria can produce reactive oxygen species due to the electron leakage at the ubiquinone site – the ubiquinone:cytochrome b region – and at the matrix side of complex I (NADH dehydrogenase.)

Superoxide radical and its reduction product $H_2O_2$ are potentially toxic compounds, and can also combine by the Haber-Weiss reaction to form the highly toxic $OH^-$. Many reports show the deleterious effects of re-active oxygen species, whose production is stimulated under water stress. Reactive oxygen species cause lipid peroxidation, and consequently membrane injuries, protein degradation and enzyme inactivation. Oxidative stress may also cause protein oxidation, with a loss of enzyme activity and the formation of protease-resistant cross-linked aggregates. Oxidatively-damaged proteins accumulate in pea leaves subjected to moderate water stress.

Overall, the production of reactive oxygen species is linear with the severity of drought stress, which leads to enhanced peroxidation of membrane lipids and degradation of nucleic acids, and both structural and functional proteins. Various organelles including chloroplasts, mitochondria and peroxisomes are the seats as well as first target of reactive oxygen species produced under drought stress.

## Drought Resistance Mechanisms

Plants respond and adapt to and survive under drought stress by the induction of various morphological, biochemical and physiological responses. Drought tolerance is defined as the ability to grow, flower and display economic yield under suboptimal water supply. Drought stress affects the water relations of plants at cellular, tissue and organ levels, causing specific as well as unspecific reactions, damage and adaptation reactions. To cope with the drought, tolerant plants initiate defense mechanisms against water deficit, which need to be investigated in further detail.

## Morphological Mechanisms

Plant drought tolerance involves changes at whole-plant, tissue, physiological and molecular levels. Manifestation of a single or a combination of inherent changes determines the ability of the plant to sustain itself under limited moisture supply.

## Escape

Escape from drought is attained through a shortened life cycle or growing season, allowing plants to reproduce before the environment becomes dry. Flowering time is an important trait related to drought adaptation, where a short life cycle can lead to drought escape. Crop duration is interactively determined by genotype and the environment and determines the ability of the crop to escape from climatic stresses including drought. Matching growth duration of plants to soil moisture availability is critical to realize high seed yield. Drought escape occurs when phenological development is successfully matched with periods of soil moisture availability, where the growing season is shorter and terminal drought stress predominates. In field-grown clones of robusta coffee, leaf shedding in response to drought stress occurred sequentially from older to younger leaves, suggesting that the more drought-sensitive the clone, the greater the extent of leaf shedding.

Time of flowering is a major trait of a crop adaptation to the environment, particularly when the growing season is restricted by terminal drought and high temperatures. Developing short-duration varieties has been an effective strategy for minimizing yield loss from terminal drought, as early maturity helps the crop to avoid the period of stress. However, yield is generally correlated with the length of crop duration under favorable growing conditions, and any decline in crop duration below the optimum would tax yield.

## Avoidance

Drought avoidance consists of mechanisms that reduce water loss from plants, due to stomatal control of transpiration, and also maintain water uptake through an extensive and prolific root system. The root characters such as biomass, length, density and depth are the main drought avoidance traits that contribute to final yield under terminal drought environments. A deep and thick root system is helpful for extracting water from considerable depths.

Glaucousness or waxy bloom on leaves helps with maintenance of high tissue water potential, and is therefore considered as a desirable trait for drought tolerance. Varying degrees of glaucousness in wheat led to increased water-use efficiency, but did not affect total water use or harvest index. Determination of leaf temperature indicated that, compared with non-glaucous leaves, glaucous leaves were 0.7°C cooler and had a lower rate of leaf senescence. These authors suggested that a 0.5°C reduction in leaf temperature for six hours per day was sufficient to extend the grain-filling period by more than three days. However, yield advantages are likely to be small as many varieties already show some degree of glaucousness.

## Phenotypic Flexibility

Plant growth is greatly affected by water deficit. At a morphological level, the shoot and root are the most affected and both are the key components of plant adaptation to drought. Plants generally limit the number and area of leaves in response to drought stress just to cut down the water budget at the cost of yield loss. Since roots are the only source to acquire water from soil, the root growth, its density, proliferation and size are key responses of plants to drought stress.

It has long been established that plants bearing small leaves are typical of xeric environments. Such plants withstand drought very well, albeit their growth rate and biomass are relatively low. Leaf pubescence is a xeromorphic trait that helps protect the leaves from excessive heat load. Hairy leaves have reduced leaf temperatures and transpiration whilst inter and intra-specific variation exists for the presence of this trait. Under high temperature and radiation stress, hairiness increases the light reflectance and minimizes water loss by increasing the boundary layer resistance to water vapor movement away from the leaf surface. Although drought stress also induces the production of trichomes on both sides of wheat leaves, they had no significant influence on boundary layer resistance.

The water content in drought-treated mature stems declined by 4% and water potential by −0.25 MPa. It is shown that active phloem supply of assimilates and associated water reserves from mature stems was the mechanism that allowed developing stems of Hylocereus undatus to maintain growth under drought conditions. Moreover, girdling the phloem of growing stems rapidly inhibited stem elongation, but secretion of sucrose-containing nectar was maintained during drought. The water potential gradient was in the wrong direction for xylem transport from mature to young growing stems and axial hydraulic conductivity was low to negligible.

Roots are the key plant organ for adaptation to drought. If tolerance is defined as the ability to maintain leaf area and growth under prolonged vegetative stage stress, the main basis of variation appears to be constitutive root system architecture that allows the maintenance of more favorable plant water status. The possession of a deep and thick root system allowed access to water deep in the soil, which was considered important in determining drought resistance in upland rice. Evidence suggests that it is quality, i.e. the distribution and structure, and not quantity of roots that determines the most efficient strategy for extracting water during the crop-growing season. The drought tolerance of tea, onion and cotton was increased by improved root growth and root functioning. Selection for a deep and extensive root system has been advocated to increase productivity of food legumes under moisture-deficit conditions as it can optimize the capacity to acquire water.

Studies carried out on the effects of alleles of the wheat shoot dwarfing genes on root-shoot dry matter partitioning and drought resistance revealed that cultivars possessing the reduced height gene 1 and reduced height gene 2 gibberellin-insensitive dwarfing genes were more susceptible to drought stress than reduced height gene 1 and reduced height gene 2 tall cultivars. The semi-dwarfing genes reduced height gene 1 and reduced height gene 2 resulted in greater root biomass at anthesis due to increased thickening of existing roots using surplus assimilates arising from the restricted stem growth. Thus, the benefit of greater assimilates available for root growth was not expressed as more extensive or deeper root growth. Differences have also been observed in the adaptive response of root distribution to soil drying.

To summarize, plants may escape drought stress by cutting short their growth duration, and avoid the stress with the maintenance of high tissue water potential either by reducing water loss from plants or improved water uptake, or both. Some plants may reduce their surface area either by leaf shedding or production of smaller leaves.

## Physiological Mechanisms

Osmotic adjustment, osmoprotection, antioxidation and a scavenging defense system have been the most important bases responsible for drought tolerance. The physiological basis of genetic variation in drought response is not clear; in part, because more intricate mechanisms have been suggested. Some of these mechanisms are described below.

Root growth and proliferation under well-watered and drought stress conditions in various rice genotypes. Different rice genotypes (Nip, sl 13, sl 34, sl 45, sl 50) were grown under continuous flooded conditions (well-watered) and 15% soil moisture contents (drought stress). The study was conducted in root boxes. The figure shows root proliferation 38 days after seeding.

## Cell and Tissue Water Conservation

Under drought stress, sensitive pea genotypes were more affected by a decline in relative water content than tolerant ones. In faba bean, determination of leaf water potential was useful for describing the drought effect, but was not suitable for discriminating tolerant from sensitive genotypes. This suggested that water potential was not the defining feature of the tolerance. Nevertheless, other studies opined that determination of leaf water status in the morning and water content in leaves in the afternoon were potentially useful for screening drought tolerance in chickpea.

Osmotic adjustment allows the cell to decrease osmotic potential and, as a consequence, increases the gradient for water influx and maintenance of turgor. Improved tissue water status may be achieved through osmotic adjustment and changes in cell wall elasticity. This is essential for maintaining physiological activity for extended periods of drought. Wild melon plant survived drought by maintaining its water content without wilting of leaves even under severe drought. Drought stress in combination with strong light led to an accumulation of high concentrations of citrulline, glutamate and arginine in leaves of wild watermelon. The accumulation of citrulline and arginine may be related to the induction of dopamine receptor interacting protein gene 1, a homologue of the acetylornithine deacetylase gene in Es-cherichia coli, where it functions to incorporate the carbon skeleton of glutamate into the urea cycle.

It has been identified that among various mechanisms, osmotic adjustment, abscisic acid and induction of dehydrins may confer tolerance against drought injuries by maintaining high tissue water potential. With the accumulation of solutes, the osmotic potential of the cell is lowered, which attracts water into the cell and helps with turgor maintenance. The maintenance of turgor despite a decrease in leaf water volume is consistent with other studies of species with elastic cell walls. Osmotic adjustment helps to maintain the cell water balance with the active accumulation of solutes in the cytoplasm, thereby minimizing the harmful effects of drought. Osmotic adjustment is an important trait in delaying dehydrative damage in water-limited environments by continued maintenance of cell turgor and physiological processes. The osmotic adjustment also facilitates a better translocation of preanthesis carbohydrate partitioning during grain filling, while high turgor maintenance leads to higher photosynthetic rate and growth.

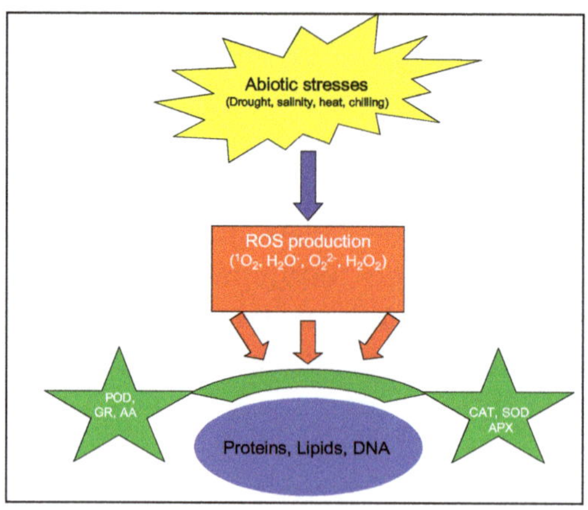


Role of antioxidant enzymes in the ROS scavenging mechanism. Exposure to abiotic stresses (including drought, chilling, salinity, etc.) leads to the generation of ROS, including singlet oxygen ($1O_2$), perhydroxyl radical ($H_2O^·$), hydroxyl radicals ($O_2^{-2}$), hydrogen peroxide ($H_2O_2$) and alkoxy radical (RO). The ROS may react with proteins, lipids and DNA, causing oxidative damage and impairing the normal functions of cells. The antioxidant defense system in the plant cell includes both enzymatic and non-enzymatic constituents. Amongst the enzymatic components are superoxide dismutase, catalase, peroxidase, ascorbate peroxidase and glutathione reductase. Upon exposure to abiotic stresses, tolerant cells activate their enzymatic antioxidant system, which then starts quenching the ROS and protecting the cell.

## Antioxidant Defense

The antioxidant defense system in the plant cell constitutes both enzymatic and non-enzymatic components. Enzymatic components include superoxide dismutase, catalase, peroxidase, ascorbate peroxidase and glutathione reductase. Non-enzymatic components contain cystein, reduced glutathione and ascorbic acid. In environmental stress tolerance, such as drought, high activities of antioxidant enzymes and high contents of non-enzymatic constituents are important.

The reactive oxygen species in plants are removed by a variety of antioxidant enzymes and lipid-soluble and water-soluble scavenging molecules; the antioxidant enzymes being the most efficient mechanisms against oxidative stress. Apart from catalase, various peroxidases and peroxiredoxins, four enzymes are involved in the ascorbate-glutathione cycle, a pathway that allows the scavenging of superoxide radicals and $H_2O_2$. These include ascorbate peroxidase, dehydroascorbate reductase, monodehydroascorbate reductase and glutathione reductase. Most of the ascorbateglutathione cycle enzymes are located in the cytosol, stroma of chloroplasts, mitochondria and peroxisomes. Ascorbate peroxidase is a key antioxidant enzyme in plants whilst glutathione reductase has a central role in maintaining the reduced glutathione pool during stress. Two glutathione reductase complementary deoxyribonucleic acids have been isolated; one type encoding the cytosolic isoforms and the other encoding glutathione reductase proteins dual-targeted to both chloroplasts and mitochondria in different plants.

Among enzymatic mechanisms, superoxide dismutase plays an important role, and catalyzes the dismutation of two molecules of superoxide into $O_2$ and $H_2O_2$; the first step in reactive oxygen species scavenging systems. Lima et al. from a study utilizing two rapidly drought-stressed clones of Coffea canephora, proposed that drought tolerance might, or at least in part, be associated with enhanced activity of antioxidant enzymes. In contrast, Pinheiro et al. did not find a link between protection against oxidative stress and drought tolerance when four clones of C. canephora were subjected to long-term drought.

Carotenoids and other compounds, such as abietane diterpenes, have received little attention despite their capacity to scavenge singlet oxygen and lipid peroxy-radicals, as well as to inhibit lipid peroxidation and superoxide generation under dehydrative forces. The transcript of some of the antioxidant genes such as glutathione reductase or ascorbate peroxidase was higher during recovery from a water deficit period and appeared to play a role in the protection of cellular machinery against damage by reactive oxygen species. A superoxide radical has a half-life of less than 1 sec and is rapidly dismutated by superoxide dismutase into $H_2O_2$, a product that is relatively stable and can be detoxified by catalase and peroxidase. These metalloenzymes constitute an important primary line of defense of cells against superoxide free radicals generated under stress conditions. Therefore, increased superoxide dismutase activity is known to confer oxidative stress tolerance.

Oxidative damage in the plant tissue is alleviated by a concerted action of both enzymatic and non-enzymatic antioxidant systems. These include β-carotenes, ascorbic acid, α-tocopherol, reduced glutathione and enzymes including superoxide dismutase, peroxidase, ascorbate peroxidase, catalase, polyphenol oxidase and glutathione reductase. Carotenes form a key part of the plant antioxidant defense system, but they are very susceptible to oxidative destruction. The β-carotene present in the chloroplasts of all green plants is exclusively bound to the core complexes of photosystem I and photosystem II. Protection against damaging effects of reactive oxygen species at this site is essential for chloroplast functioning. Here, β-carotene, in addition to functioning as an accessory pigment, acts as an effective antioxidant and plays a unique role in protecting photochemical processes and sustaining them. A major protective role of β-carotene in photosynthetic tissue may be through direct quenching of triplet chlorophyll, which prevents the generation of singlet oxygen and protects from oxidative damage.

## Cell Membrane Stability

Biological membranes are the first target of many abiotic stresses. It is generally accepted that the maintenance of integrity and stability of membranes under water stress is a major component of drought tolerance in plants. Cell membrane stability, reciprocal to cell membrane injury, is a physiological index widely used for the evaluation of drought tolerance. Moreover, it is a genetically related phenomenon since quantitative trait loci for this have been mapped in drought-stressed rice at different growth stages. Dhanda et al. showed that membrane stability of the leaf segment was the most important trait to screen the germplasm for drought tolerance.

Cell membrane stability declined rapidly in Kentucky bluegrass exposed to drought and heat stress simultaneously. In a study on maize, K nutrition improved the drought tolerance, mainly due to improved cell membrane stability. Tolerance to drought evaluated as increase in cell membrane stability under water deficit conditions was differentiated between cultivars and correlated well with a reduction in relative growth rate under stress. In holm oak (Quercus ilex) seedlings, hardening increased drought tolerance primarily by reducing osmotic potential and stomatal regulation, improved new root growth capacity and enhanced cell membrane stability. Among treated seedlings, the greatest response occurred in seedlings subjected to moderate hardening. Variation in cell membrane stability, stomatal regulation and root growth capacity was negatively related to osmotic adjustment.

The causes of membrane disruption are unknown; not withstanding, a decrease in cellular volume causes crowding and increases the viscosity of cytoplasmic components.

This increases the chances of molecular interactions that can cause protein denaturation and membrane fusion. For model membrane and protein systems, a broad range of compounds have been identified that can prevent such adverse molecular interactions. Some of these are proline, glutamate, glycinebetaine, carnitine, mannitol, sorbitol, fructans, polyols, trehalose, sucrose and oligosaccharides. Another possibility of ion leakage from the cell may be due to thermal-induced inhibition of membrane-bound enzymes responsible for maintaining chemical gradients in the cell. Arabidopsis leaf membranes appeared to be very resistant to water deficit, as shown by their capacity to maintain polar lipid contents and the stability of their composition under severe drought.

## Plant Growth Regulators

Plant growth regulators, when applied externally, and phytohormones, when produced internally, are substances that influence physiological processes of plants at very low concentrations. Both these terms have been used interchangeably, particularly when referring to auxins, gibberellins, cytokinins, ethylene and abscisic acid. Under drought, endogenous contents of auxins, gibberellins and cytokinin usually decrease, while those of abscisic acid and ethylene increase. Nevertheless, phytohormones play vital roles in drought tolerance of plants.

Auxins induce new root formation by breaking root apical dominance induced by cytokinins. As a prolific root system is vital for drought tolerance, auxins have an indirect but key role in this regard. Drought stress limits the production of endogenous auxins, usually when contents of abscisic acid and ethylene increase. Nevertheless, exogenous application of indole-3-yl-acetic acid enhanced net photosynthesis and stomatal conductance in cotton. Indole-3-butyric acid is a naturally occurring auxin. Drought stress and abscisic acid application enhance indole-3-butyric acid synthesis in maize. Recently, it was revealed that Indole-3-butyric acid synthetase from Arabidopsis is also drought-inducible. Experiments with indole-3-yl-acetic acid and ethylene glycol tetra-acetic acid suggested that calcium and auxin participate in signaling mechanisms of drought-induced proline accumulation.

Drought rhizogenesis is an adaptive strategy that occurs during progressive drought stress and is reported from Brassicaceae and related families by the formation of short, tuberized, hairless roots. These roots are capable of withstanding a prolonged drought period and give rise to a new functional root system upon rehydration. The drought rhizogenesis was highly increased in the gibberrelic acid biosynthetic mutant ga5, suggesting that some gibberrelic acids might also participate in this process.

Abscisic acid is a growth inhibitor and produced under a wide variety of environmental stresses, including drought. All plants respond to drought and many other stresses by accumulating abscisic acid. Abscisic acid is ubiquitous in all flowering plants and is generally recognized as a stress hormone that regulates gene expression and acts as a signal for the initiation of processes involved in adaptation to drought and other environmental stresses. It has been proposed that abscisic acid and cytokinin have opposite roles in drought stress. Increase in abscisic acid and decline in cytokinins levels favor stomatal closure and limit water loss through transpiration under water stress. When plants wilt, abscisic acid levels typically rise as a result of increased synthesis. Increased abscisic acid concentration leads to many changes in development, physiology and growth. Abscisic acid alters the relative growth rates of various plant parts such as increase in the root-to-shoot dry weight ratio, inhibition of leaf area development and production of prolific and deeper roots. It triggers the occurrence of a complex series of events leading to stomatal closure, which is an important water-conservation response. In a study on genetic variation for abscisic acid accumulation in rice, a consistent negative relationship between the ability of detached and partially dehydrated leaves to accumulate abscisic acid and leaf weight was established. By its effect in closing stomata, abscisic acid can control the rate of transpiration and, to some extent, may be involved in the mechanism conferring drought tolerance in plants.

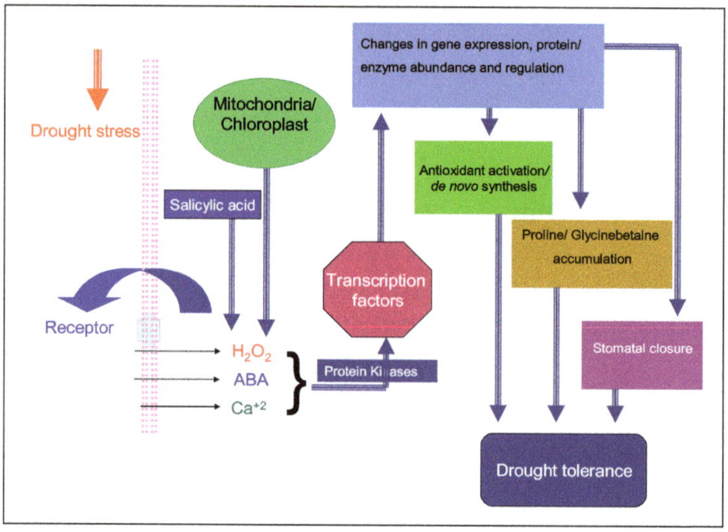


Proposed cellular events and signaling cascades in a plant cell responding to drought stress. Drought stress is perceived by an unknown mechanism, which then activates the signaling cascades, plausibly by abcissic acid (ABA), hydrogen peroxide ($H_2O_2$) and calcium ($Ca^{+2}$). These cascades then activate the synthesis of specific protein kinases which activate more downstream responses such as changes in gene expression. The response to these signaling cascades also results in changes in plant metabolism including activation and synthesis of antioxidants, synthesis and accumulation of osmoprotectants and solutes, and stomatal closure under acute drought stress.

Abscisic acid induces expression of various water stress-related genes. In a recent study, Zhang et al. reported a regulatory role of telomeric repeat binding factor gene 1 in abscisic acid sensitivity and drought response during seedling development. Bray suggested the existence of abscisic acid-dependent and abscisic acid-independent transduction cascades and pathways to act as a signal of drought stress and the expression of specific water stress-induced genes. Abscisic acid produces such changes that confer an ability to maintain cellular turgor to withstand dehydrative forces.

Ethylene has long been considered a growth inhibitory hormone, although it is involved in environmentally driven growth inhibition and stimulation. The response of cereals to drought includes loss of leaf function and premature onset of senescence in older leaves. Ethylene may serve to regulate leaf performance throughout its lifespan as well as to determine the onset of natural senescence and mediate drought-induced senescence. Recent studies suggest that growth promotion is a common feature in ethylene responses. To escape this adversity, plants can optimize growth and tolerate abiotic stresses such as drought, and this response also involves ethylene synthesis.

Among the other endogenously produced growth regulating factors, the role of salicylic acid in the induction of tolerance against several abiotic stresses has been emphasized recently. In the case of drought tolerance, the role of endogenously produced salicylic acid is still enigmatic. Salicylic acid potentiates the generation of reactive oxygen species in photosynthetic tissues of Arabidopsis thaliana during osmotic stress.

Polyamines are known to have profound influence on plant growth and development. Being cationic, polyamines can associate with anionic components of the membrane, such as phospholipids, thereby protecting the lipid bilayer from deteriorating effects of stress. There has been a growing interest in the study of polyamine participation in the defense reaction of plants against environmental stresses and extensive research efforts have been made in the last two decades. Many genes for enzymes involved in polyamine metabolism have been cloned from several species, and their expression under several stress conditions has been analyzed. For example, the apple spermidine synthase gene when overexpressed encodes high levels of spermidine synthase, which substantially improves abiotic stress tolerance including drought.

Among various polyamines, a rise in the putrescence level is generally due to an enhanced arginine decarboxylase activity. Compared with sensitive plants, stress-tolerant plants generally have a greater capacity to synthesize polyamines in response to stress, resulting in a two-to three fold rise in endogenous polyamine levels over the unstressed ones. Recent studies suggested that rice has a great capacity to enhance polyamine biosynthesis, particularly spermidine and spermine in free form and putrescence in insoluble-conjugated form, in leaves earlier in response to drought stress. This was considered as an important physiological trait of drought tolerance in rice.

Among various polyamines, a rise in the putrescence level is generally due to an enhanced arginine decarboxylase activity. Compared with sensitive plants, stress-tolerant plants generally have a greater capacity to synthesize polyamines in response to stress, resulting in a twoto three fold rise in endogenous polyamine levels over the unstressed ones. Recent studies suggested that rice has a great capacity to enhance polyamine biosynthesis, particularly spermidine and spermine in free form and putrescence in insoluble-conjugated form, in leaves earlier in response to drought stress. This was considered as an important physiological trait of drought tolerance in rice.

## Compatible Solutes and Osmotic Adjustment

One of the most common stress tolerance strategies in plants is the overproduction of different types of compatible organic solutes. Compatible solutes are low-molecular-weight, highly soluble compounds that are usually nontoxic even at high cytosolic concentrations. Generally they protect plants from stress through different means such as contribution towards osmotic adjustment, detoxification of reactive oxygen species, stabilization of membranes, and native structures of enzymes and proteins.

Osmotic adjustment is a mechanism to maintain water relations under osmotic stress. It involves the accumulation of a range of osmotically active molecules/ions including soluble sugars, sugar alcohols, proline, glycinebetaine, organic acids, calcium, potassium, chloride ions, etc. Under water deficit and as a result of solute accumulation, the osmotic potential of the cell is lowered, which attracts water into the cell and helps with the maintenance of turgor. By means of osmotic adjustment, the organelles and cytoplasmic activities take place at about a normal pace and help plants to perform better in terms of growth, photosynthesis and assimilate partitioning to grain filling. As a mechanism, osmotic adjustment has been suggested as an important trait in postponing the dehydration stress in water-scarce environments. Variation in osmotic adjustment among chickpea cultivars in response to soil drought has been observed, and seed yield of chickpea was correlated with the degree of osmotic adjustment when grown under a line-source irrigation system in the field. Contrarily, Serraj and Sinclair found no yield advantage from osmotic adjustment in any crop. Nevertheless, further investigations are imperative to establish this controversy.

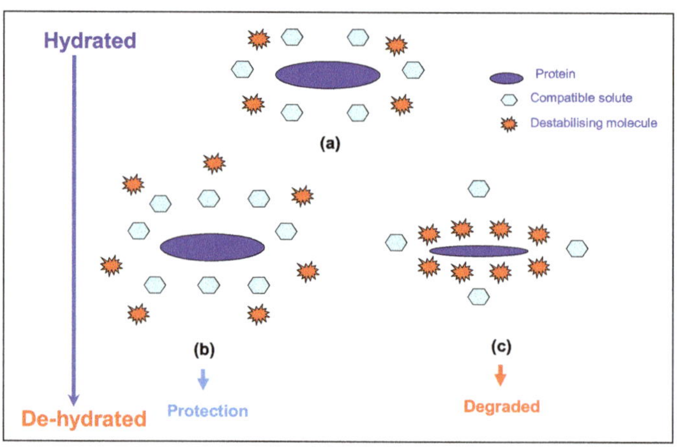


Role of compatible solutes in drought tolerance. In the hydrated state, the presence of water reduces the interaction of destabilizing molecules (a), in tolerant cells the synthesis of compatible solutes preferentially excludes the binding of destabilizing molecules and stabilizes native protein conformation (b) and in sensitive cells the lack of compatible solutes results in the preferential binding of destabilizing molecules to the protein surface, leading to degradation (c).

Osmotic adjustment is accomplished with the accumulation of compatible solutes. Of these, proline is one amongst the most important cytosolutes and its free accumulation is a widespread response of higher plants, algae, animals and bacteria to low water potential. Its synthesis in leaves at low water potential is caused by a combination of increased biosynthesis and slow oxidation in mitochondria. Despite some controversy, many physiological roles have been assigned to free proline including stabilization of macromolecules, a sink for excess reductant and a store of carbon and nitrogen for use after relief of water deficit. Proline contents were increased under drought stress in pea cultivars. Drought-tolerant petunia (Petunia hybrida) varieties were reported to accumulate free proline under drought that acted as an osmoprotectant and induced drought tolerance.

Glycinebetaine (N, N, N-trimethyl glycine) is one of the most extensively studied quaternary ammonium compounds and compatible solutes in plants, animals and bacteria. Many studies demonstrate that glycinebetaine plays an important role in enhancing plant tolerance under a range of abiotic stresses including drought. The introduction of genes synthesizing glycinebetaine into non-accumulators of glycinebetaine proved to be effective in increasing tolerance to various abiotic stresses. Naidu et al. reported that cotton cultivars adapted to water stress conditions accumulated higher glycinebetaine than the non-adapted ones under drought. In addition to direct protective roles of glycinebetaine either through positive effects on enzyme and membrane integrity or as an osmoprotectant, glycinebetaine may also protect cells from environmental stresses indirectly by participating in signal transduction pathways.

Citrulline, named after Citrullus; a Latin name of watermelon, from which it was isolated, is an amino acid. Although not built into proteins during their synthesis, and not encoded by a nuclear gene, several proteins are known to contain citrulline. Wild watermelon (Citrullus lanatus) has the ability to adapt to severe drought stress despite carrying out normal C3-type photosynthesis, which seem to be correlated with citrulline accumulation. Wild watermelon primarily accumulated citrulline followed by glutamate and arginine, in place of proline and glycinebetaine. Yokota et al. reported a higher citrulline accumulation in the wild watermelon leaves assuming that citrulline is located only in the cytosol and constitutes 5% of the total volume of the mesophyll cells. Citrulline is a novel and the most effective $OH^-$ scavenger among compatible solutes examined so far. Moreover, it can effectively protect DNA and enzymes from oxidative injuries.

Rapid accumulation of the non-protein amino acid γ-aminobutyric acid was identified in plant tissues upon exposure to stress many years ago. γ-aminobutyric acid acts as a zwitterion, exists in free form, and has a flexible molecule that can assume several conformations in solution, including a cyclic structure that is similar to proline. At physiological pH, γ-aminobutyric acid is highly water-soluble, and may function as a signaling molecule in higher plants under stress. The physiological roles of γ-aminobutyric acid in drought tolerance entail osmotic regulation, detoxication of reactive oxygen radicals, conversion of putrescine into proline and intracellular signal transduction.

Drought stress initiates a signal transduction pathway, in which increased cytosolic $Ca^{2+}$ activates $Ca^{2+}$/calmodulin-dependent glutamate decarboxylase activity, leading to γ-aminobutyric acid synthesis. Elevated $H^+$ and substrate levels can also stimulate glutamate decarboxylase activity, leading primarily to γ-aminobutyric acid accumulation. Experimental evidence supports the involvement of γ-aminobutyric acid in pH regulation, nitrogen storage, plant development and defense, as well as a compatible osmolyte and an alternative pathway for glutamate utilization. After drought stress the content of proline was more than 50% and at the end of recovery the γ-aminobutyric acid content reached 27%.

Trehalose is a non-reducing disaccharide of glucose that functions as a compatible solute in the stabilization of biological structures under abiotic stress. In nature, trehalose is biosynthesized as a stress response by a variety of organisms including bacteria, fungi, algae, insects, invertebrates and lower plants. Capacity to produce trehalose, earlier thought to be absent from higher plants, has now been reported to accumulate in high amounts in some drought-tolerant ferns, the resurrection plant Selaginella lepidophylla and desiccation-tolerant angiosperm Myrothamnus flabellifolia. The presence of low amounts of trehalose was demonstrated even in tobacco and many higher plants. Its metabolism may be channelized to enhance drought tolerance in plants. Physiological roles of trehalose include efficient stabilization of dehydrated enzymes, proteins and lipid membranes, as well as protection of biological structures under desiccation stress rather than regulating water potential. Karim et al. reported that enhanced drought tolerance by trehalose depends on improved water status and expression of heterologous trehalose biosynthesis genes during Arabidopsis root development.

At a molecular level, exogenously applied trehalose may trigger the abscisic acid-insensitive 4 gene expression but decrease sucrose induction, providing a possible molecular mechanism for the trehalose effect on plant gene expression and growth. Trehalose-accumulating organisms produce this sugar in a two-step process by the action of the enzymes trehalose-6-phosphate synthase and trehalose-6-phosphate phosphatase when exposed to stress. Improved drought tolerance has been reported in the transgenic plants overproducing trehalose-6-phosphate synthase in spite of minute accumulation of trehalose. In fact, plants can withstand drought stress by conserving cell and tissue water principally by osmatic adjustment, maintenance of the antioxidant defense system for the scavenging of reactive oxygen species, and keeping the cell membranes stabilized. Plant growth regulators and polyamines, γ-aminobutyric acid, free amino acids and sugars also play a vital role in drought tolerance by scavenging the reactive oxygen species, stomatal regulation and protection of vital macro-molecules, and maintenance of thc ccll watcr balancc.

# MINERAL STRESS

Mineral stress, defined as sub-optimal availability of essential nutrients or toxicity of nutrient or non-nutrient minerals (especially Al, Na, Cl, Mn, and other heavy metals), is a primary constraint to plant growth over the majority of the earth's land surface. Natural ecosystems, managed forests and rangelands, and agriculture in less developed countries (LDCs) are largely characterized by multiple mineral stresses. This being the case, we will not be able to understand or predict ecosystem responses to global change variables without understanding how these variables interact with mineral stresses. Unfortunately, apart from N, this topic has received relatively little research attention. Some edaphic stresses of global importance, plant adaptations to edaphic stress (focusing on organismic processes), and ideas about how these adaptations might be affected by global change.

Table: Mention of an element in an article often resulted from routine tissue analysis rather than specific study of the interaction of elevated $CO_2$ with mineral stress.

| Key words | Number of journal articles |
|---|---|
| Elevated $CO_2$, plants | 1658 |
| Elevated $CO_2$, plants + N | 685 |
| Elevated $CO_2$, plants + P | 82 |
| Elevated $CO_2$, plants + K | 10 |
| Elevated $CO_2$, plants + Ca | 15 |
| Elevated $CO_2$, plants + Mg | 7 |
| Elevated $CO_2$, plants + Fe | 2 |
| Elevated $CO_2$, plants + Mn | 3 |
| Elevated $CO_2$, plants + Cu | 0 |
| Elevated $CO_2$, plants + Zn | 1 |
| Elevated $CO_2$, plants + B | 1 |
| Elevated $CO_2$, plants + Al | 2 |
| Elevated $CO_2$, plants + Si | 1 |

## Mineral Stresses of Global Importance

Edaphic stresses include physical factors such as soil shallowness, poor drainage, and poor water retention, as well as chemical factors such as sub-optimal availability of mineral nutrients, acidity, alkalinity, salinity, and metal toxicity. Chemical factors are more important limitations to plant growth than physical factors on a global basis.

## Nutrient Deficiency

Nutrient deficiency is the norm in native soils. The most inherently fertile soils such as Mollisols generally support dense plant communities characterized by intense nutrient competition, so even in these soils plants can be nutrient deficient. Sub-optimal availability of N and P are nearly universal. Imbalances of base cations (K, Ca, and Mg) and micronutrient deficiencies affect large areas. Intensive agriculture in developed nations has alleviated nutrient deficiencies, but the majority of world agriculture occurs with low fertility inputs on soils with poor availability of P and other nutrients.

## Soil Acidity

Soil acidity (i.e., pH below 5.5) is characteristic of weathered soils of the tropics and subtropics, and many temperate forest systems. These soils, including most Oxisols and Ultisols, as well as many Spodosols and Inceptisols, support most of the global vegetation as well as much of the earth's population. Acid soils are characterized by Al toxicity (below pH 5), P deficiency, poor availability of bases, especially Ca, and Mn toxicity. Negative effects of global change on these systems would have disproportionate effects on human welfare, since Third World agriculture (the single largest human occupation) is characterized by poor access to soil fertility amendments.

## Soil Alkalinity

Soil alkalinity (i.e., pH above 7.5) is characteristic of arid and semi-arid regions, and highly calcareous soils. These soils, including many Aridisols and some Entisols, are extensive but usually too dry to support dense vegetation. Alkaline soils are characterized by poor availability of P and micronutrient transition metals (Fe, Cu, Mn, and Zn).

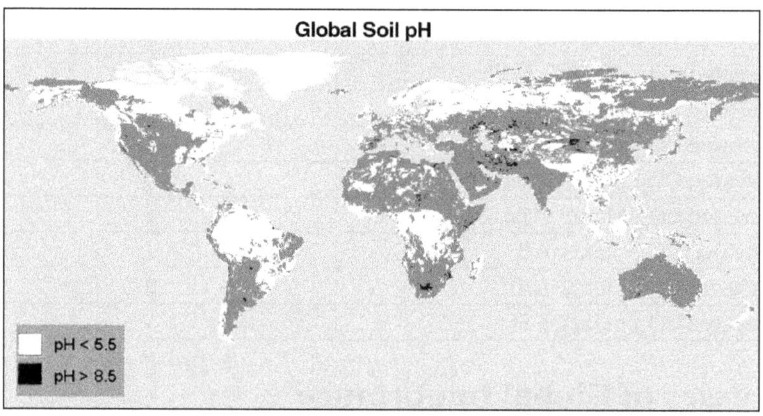


Distribution of soils having acidic (<5.5) and alkaline (>8.5) pH based on interpretation of data from the FAO soils database. Distribution of acidic (pH < 5.5, white area) and alkaline (pH >8.5, black area) soils.

## Salinity

Salinity (i.e., electrical conductivity greater than 4 dS $m^{-1}$) is characteristic of arid and semi-arid regions, and irrigated agriculture. Salinity stress is an important constraint to world agriculture, affecting 7% of the land surface of the earth and over 50% of irrigated land. Saline soils present an array of problems for plant growth, including poor water availability, toxic amounts of Na and Cl, and low availability of Ca.

## Metal Toxicity

Metal toxicity is most commonly found as Al and Mn toxicity in acid soils, although localized regions have excessive amounts of heavy metals (Pb, Cd, Cu, Zn, Mn) due to unusual soil parent material (e.g., ultramafic rock) or industrial activity. Manganese toxicity can be found in volcanic soils (Andosols) even at neutral pH. Volcanic soils typically support dense human populations such as the Pacific rim and Africa's Rift Valley.

Table: Soil pH ranges in relation to land use. Land use categories are simplified from the Anderson Land Use/Land Cover classification of the globe.

| Land use | Percentage of total area | Land use within soil pH range | | | | |
|---|---|---|---|---|---|---|
| | | Percentage with pH < 4.5 | Percentage with pH > 4.5 to <5.5 | Percentage with pH > 5.5 to <7.2 | Percentage with pH > 7.2 to < 8.5 | Percentage with pH > 8.5 |
| Agriculture | 19.6 | 8.6 | 16.9 | 22.7 | 23.2 | 13.3 |
| Forest | 26.8 | 43.7 | 47.3 | 18.9 | 3.8 | 1.9 |
| Grass/ savanna | 41.8 | 38.2 | 20.6 | 44.4 | 71.4 | 84.55 |
| Tundra | 11.9 | 9.5 | 15.2 | 14.0 | 1.6 | 0.2 |

The total area used in this assessment was 199,492,637 $km^2$. Water and some non-soil areas are not included in the total area.

Humans have alleviated the inherent edaphic limitations of native soils used for intensive agriculture in developed nations, but in most areas of the globe, human activity has significantly degraded soil fertility. A large portion of cropland in developing countries is degraded. Soil erosion is a massive problem in the humid tropics, deforestation has degraded the fertility of many tropical forest soils, acid precipitation has accelerated base leaching and forest decline in Northern Europe and the Eastern US, crop irrigation without adequate drainage and leaching has created serious salinity problems in semi-arid regions, especially in the former Soviet Republics, the Mideast and the Western US, rising water tables have salinized croplands in Australia, and poor fertilizer policies have contributed to soil acidification in China. In general, these trends are worsening, driven by the pressure of burgeoning human populations in less developed countries and by excessive resource consumption in developed countries.

## Plant Adaptations to Edaphic Stress

Plants display a remarkable array of physiological adaptations to edaphic stress. Some traits permit conservative use of all nutrients such as slow growth, while other traits represent adaptations to specific stresses such as the induction of organic acid exudation to protect root tips exposed to Al stress, the induction of strategy 1 and strategy 2 mechanisms to mobilize rhizosphere Fe in response to Fe deficiency, and the regulation of root architecture to optimize soil exploration in response to P deficiency. Root traits are particularly important adaptations to edaphic stress, since roots explore the soil, acquire nutrients and water, and interact with symbionts and soil biota.

Adaptations to edaphic stress, especially root adaptations, often entail ecological or physiological trade-offs. For example, in annual crop plants, low P availability induces alterations in root architecture that enhance topsoil foraging, since the topsoil is the region of greatest P availability. A tradeoff to enhanced topsoil foraging is greater sensitivity to drought stress. A tradeoff to increased specific root length, a common response to nutrient stress, is reduced root lifespan. A tradeoff with greater root growth may be reduced photosynthate allocation to shoot growth and reproduction. Although such tradeoffs are poorly understood, they are likely to have important ecophysiological repercussions on plant fitness in future environmental scenarios.

There are two general conceptual models of the response of individual plants to resource availability: the 'Law of the Minimum', which postulates that overall plant growth is limited by the single most limiting resource, and the Microeconomic Paradigm, that postulates that plants are optimizing entities that allocate internal resources so that no single external resource is limiting. Increasing atmospheric $CO_2$ is reducing the importance of carbon as a limiting external resource. If the Law of the Minimum is correct, this will have no effect on the growth of the majority of plants on earth, which are faced with edaphic constraints and nutrient limitations. In contrast, the Microeconomic Paradigm would predict that increased carbon availability will permit (for example) greater root growth, alleviation of nutrient stresses, and greater overall plant growth. A recent empirical study of growth responses to multiple nutrient limitations found that in some cases plants followed the Law of the Minimum, in other cases the Microeconomic Paradigm, and in some cases showed complex behavior that followed neither model. These models must be reconciled or improved if we are to predict plant response to global change.

An important feature of plant adaptations to edaphic stress is that they typically display substantial interspecific and intraspecific variation, as well as substantial phenotypic plasticity. In crop plants, it is not uncommon to find differences of 200–300% in the ability to tolerate edaphic stress, and concomitant large variation in specific traits such as root hairs, root architecture, and root exudates. This variation will facilitate natural and human selection for genotypes adapted to future environments.

## Potential Interactions of Global Change with Mineral Stress

Global change variables (GCVs) of interest here include elevated atmospheric $CO_2$, ozone, and UV radiation, and changes in spatial and temporal patterns of precipitation, temperature, humidity, and photosynthetically active radiation.

Table below lists potential interactions of GCVs with mineral stresses. Some interactions shown are supported by experimental data, but many are supported only by theory, inference, or a few research reports.

## Transpiration-driven Mass Flow

Acquisition of water-soluble nutrients such as nitrate, sulfate, Ca, Mg, and Si is primarily determined by mass flow of soil water to roots as a result of shoot transpiration. GCVs that affect transpiration should, therefore, also affect the acquisition of these nutrients. Elevated $CO_2$ normally reduces leaf level transpiration in C3 plants, and appears to reduce stand level transpiration in annual crops, and has significant but more muted effects on transpiration of forest canopies. This interaction may be particularly important for Ca and Si, two nutrients that are especially dependent on transpiration in most plant species. Elevated $CO_2$ significantly reduced Ca uptake in wheat (Triticum aestivum L.) plants. Low Ca availability is prevalent in acid soils that encompass most terrestrial forests. Logging and acid deposition have reduced Ca pools in temperate forests of Northeastern America and Western Europe, a factor associated with forest decline. Further reductions in Ca transport by elevated $CO_2$, if not offset by higher temperatures and longer warm seasons, could be detrimental. Silicon is an essential element for monocots and gymnosperms, and in dicots, it is an important component of plant resistance to insect and fungal attack. Effects of GCVs on Si transport to leaves are unknown but could have important ecological consequences.

## Root Growth and Architecture

Elevated atmospheric $CO_2$ generally increases plant carbon allocation to belowground processes, including root growth, respiration, and exudation. Belowground effects of elevated $CO_2$ could have positive or negative feedbacks on plant productivity depending on nutrient availability and other belowground processes, including sequestration of carbon into soil pools. Although data are scarce, several studies indicate that elevated atmospheric $CO_2$ may have specific effects on the architecture of roots as well as the extent and type of ectomycorrhizal symbioses, and potentially with root lifespan. Although the large number of possible interactions between climate change variables and edaphic constraints makes generalizations difficult, Kubiske and colleagues suggested that some of the effect of elevated $CO_2$ on root activity could be due to enhanced photosynthesis in the lower portion of tree canopies. Pritchard and Rogers proposed that direct and indirect signals arising from greater carbohydrate availability in plants grown with elevated $CO_2$ stimulate lateral root branching, which in turn leads to more shallow, highly branched root architectures. Whether such changes have a positive or negative impact on plant growth will depend, of course, on many specific features of the system, including the relative importance of nutrients that are enriched in surface horizons, such as phosphorus, versus those that may be found at greater depth, such as nitrogen, and whether greater root densities in a specific soil horizon result in greater soil exploration or merely greater inter-root competition. The amount of nutrients obtained by specific root architecture may be more important than the efficiency with which nutrients are acquired, especially in a high $CO_2$ world in which carbon-based currencies are relatively devalued.

Table: Some potential interactions of global change variables with mineral stress.

| Process | Global change variables | Interaction with mineral stress |
|---|---|---|
| Transpiration-driven mass flow | $CO_2$, RH, precipitation, light, temperature | $NO_3$, $SO_4$, Ca, Mg, and Si |
| Root growth and architecture | $CO_2$, precipitation, soil temperature | All nutrients, esp. P and K |
| Mycorrhizas | $CO_2$?, temperature? | P, Zn (VAM) N (ectomycorrhizas) |
| Root exudates | $CO_2$? | Al toxicity, metal uptake, and P |
| Biological N fixation | $CO_2$ | N |
| Tissue dilution | $CO_2$ | Stress responses, nutrient cycling, and nutritive value |
| Metal photochemistry | Light, UV, temperature, and precipitation | Transition metal deficiency and Mn toxicity |
| Growth responses | $CO_2$ | N and P: "no clear pattern" |

In order to predict how elevated $CO_2$ may affect root architecture, and therefore nutrient acquisition in various ecosystems, we will need a better understanding of the physiological mechanisms linking photosynthesis and root architecture, such as that proposed by Pritchard and Rogers. This is challenging, since we know very little about the physiological mechanisms that regulate root architecture in any circumstance. The dynamic responses of root architecture to the availability of nutrients and water, soil physical conditions and microbes will interact with direct effects of elevated $CO_2$, as well as indirect effects of climate change parameters on processes such as soil water status and nutrient release from plant litter. For example, is the shallow root proliferation discussed by Pritchard and Rogers the result of plant carbohydrate signals on lateral meristems or rather the result of moister epipedons because of reduced stomatal transpiration. In order to understand how root architecture will respond to a changing climate, we must therefore understand a great deal more about the regulation of root architecture in general. Genotypic variation for root architecture within populations will permit genetic drift towards architectures that are more adaptive to the climate as it changes, so analysis of current genotypes may not permit entirely accurate predictions about the response of future plant populations.

## Mycorrhizal Symbioses

Mycorrhizal symbioses are nearly ubiquitous in plants and play important roles in mobilization of nutrients from organic matter (ectomycorrhizas and ericoid mycorrhizas) and acquisition of diffusion limited nutrients, largely P and Zn (arbuscular mycorrhizas). The effects of elevated $CO_2$ and temperature on arbuscular mycorrhizas appears to be mainly indirect, i.e., mediated by plant growth responses, rather than direct. Numerous studies show increased resource allocation to mycorrhizal roots in plants exposed to elevated $CO_2$ this may reflect the relative importance of edaphic constraints to plant growth in most ecosystems. Elevated $CO_2$ appears to have complex effects on ectomycorrhizal communities in forests, with uncertain effects for tree health.

## Root Exudates

About half of belowground carbon allocation in plants is deposited in the rhizosphere. The majority of this material consists of dead root tissue, but compounds exuded by living cells, including mucilage, organic acids, phosphatases, phytosiderophores, and protons, are important in protecting growing roots from Al stress and mobilizing relatively insoluble nutrients such as P and Fe. Interactions of GCVs with root exudates should, therefore, have important consequences for plant growth in acid and alkaline soils. Only a few studies have examined this interaction, with conflicting results. For example, in some studies elevated $CO_2$ caused no effect on root exudates, while in others it decreased root exudates but altered exudate composition. Light, temperature, and other variables affecting photosynthesis have complex effects on root exudates, as might be expected. Elevated $CO_2$ increases rhizosphere respiration, but it is not clear whether this is due to increased exudates per unit root surface area or simply increased root growth. The complex interplay of root exudates with root photosynthate supply, root growth and architecture, and the rhizosphere environment make this a challenging research topic that nonetheless deserves further study because of its importance for plant adaptation to acid soils.

## Biological Nitrogen Fixation

Plant taxa capable of symbiotic nitrogen fixation are more responsive to elevated $CO_2$ than nonfixing taxa because they can directly utilize additional photosynthate to alleviate N limitation. The stimulatory effect of elevated $CO_2$ on legumes has been observed in mixed stands, such as legume–grass pastures, as well as soybean (Glycine max (L.) Merr.) monocultures. Phosphorus plays a key role in legume response to $CO_2$, since legume nodules have a high P requirement.

## Tissue Dilution

Elevated $CO_2$ can increase carbon gain in C3 plants, which can dilute tissue nutrient concentration. For plants grown at double the current $CO_2$ concentration, the C:N ratio increases on average about 15%. Dilution effects for P are uncertain due to a small number of observations and variable results. Elevated $CO_2$ reduced the content of Ca, S, Mg, and Zn in wheat plants as a result of dilution caused by tissue carbohydrate accumulation, combined with transpiration-driven reductions in Ca uptake. Reduced nutrient concentration in plant tissue could trigger plant stress responses, although this may not be generally true since bulk nutrient concentration may be reduced by starch accumulation and tissue density effects, which may not directly affect cytoplasmic nutrient availability in chlorenchyma and other metabolically active tissue. Nutrient dilution is more likely to have important consequences at the ecosystem level, such as slower litter decay, although this effect appears to be slight and dietary effects on leaf-feeding insects and other herbivores, including humans. Interestingly, deleterious effects of elevated $CO_2$ on wheat protein content were pronounced at low N but slight at high N.

## Metal Photochemistry

Toxic levels of reactive oxygen species can be formed in chloroplasts in certain conditions. The role of metals (Mn, Mg, Fe, and Cu) in the light reactions of photosynthesis, as well as in antioxidant enzyme systems (Zn, Cu, and Mn in SOD, Fe in catalase) makes them important in both the generation and the detoxification of reactive oxygen species generated in chloroplasts. A number of GCVs can increase the formation of reactive oxygen species, including ozone, high light, UV radiation, temperature extremes, and drought. Plants suffering sub-optimal availability of Mg and transition metals because of high soil pH, base imbalances, Al toxicity, and Mn toxicity may be more sensitive to global change than healthy plants. Manganese toxicity is discussed below as a specific example of the role of metal photochemistry in mediating plant response to light intensity.

## Growth Responses

A principal concern in predicting plant response to global change is whether the growth promotive effects of greater atmospheric $CO_2$ concentration and, in some regions, greater N deposition or precipitation will be nullified by other limitations to plant growth, such as nutrient availability. To this end, a number of studies have addressed the interaction of elevated $CO_2$ with N availability, and a smaller number of studies have addressed the interaction of elevated $CO_2$ with P availability. Recent reviews of this literature conclude that although nutrient stress tended to reduce growth stimulation by elevated $CO_2$ in both herbaceous and woody species, results show "no clear pattern": Interactions between elevated $CO_2$ and nutrient stress ranged from antagonist to neutral to synergistic. The complexity of growth responses to resource interactions is consistent with the view that such interactions involve complex biology that varies among organisms, environments, and specific resources, and is not readily predicted from simple conceptual models.

## Effects of Temperature on Plant-nutrient Interactions in a Changing Environment

Relatively little is known about the influence of temperature on plant–nutrient interactions in a changing environment. In an analysis of the available literature in 2001, low-soil nutrition and sub-optimal temperature had the greatest influence on reducing the growth-stimulating effects of elevated $CO_2$. Soil temperature has marked effects on root characteristics (growth, morphology, longevity, respiration, and membrane fluidity), which influence plant nutrient acquisition efficiency. Studies generally suggest that soil warming enhances nutrient uptake. However, three studies showed no interactive effect of elevated $CO_2$ and temperature on P and N acquisition suggesting that higher temperature does not modulate the nutrient dilution effect often observed in plants exposed to elevated $CO_2$. Several phenolic compounds in willow leaves were decreased in response to higher temperature. Such effects on leaf secondary chemistry could influence nutrient cycling in soils by increasing litter decomposition rates, which are often slowed as C/N ratios increase under elevated $CO_2$.

## Effects of Light on Plant–nutrient Interactions in a Changing Environment

Increases in UV radiation and visible light increase foliar phenolic content. Altered secondary chemistry can decrease litter decomposition rates, affecting nutrient cycling dynamics in the soil. However, the effects of UV radiation on litter decomposition are highly variable in the literature.

High visible light intensity and excessive UV light can trigger photoinhibition leading to photo-oxidative stress, particularly in the presence of other environmental stresses. Foliar nutrient imbalances including nutrient deficiencies and metal toxicity interact strongly with high light to induce photo-oxidative stress, which can have significant negative effects on plant productivity. Because of its influence on primary plant metabolism, elevated $CO_2$ in theory could have important influences on the redox balance in plant cells. No clear patterns have emerged with regard to how antioxidant systems respond to elevated $CO_2$. Depending on species and environmental conditions, elevated $CO_2$ has been shown to both induce and repress antioxidant systems in plants.

## Example of Manganese Toxicity

Manganese (Mn) toxicity, an important mineral stress in both natural and agricultural systems, highlights the complexity of the interactions of climate change variables with mineral stress. One mechanism of Mn phytotoxicity is photo-oxidation. Manganese toxicity could generate photo-oxidative stress through several mechanisms, including (1) direct generation of reactive oxygen from Mn ions in the presence of light through the Fenton reaction, (2) disabling of antioxidant systems reliant on Fe (such as catalase), Cu and Zn (such as forms of SOD) by interference with metal uptake, transport, and metabolism, (3) disruption of normal electron flow from PSII to PSI by antagonism with electron acceptors with metal cofactors, such as plastocyanin, (4) disruption of normal energy dissipation via Rubisco inactivation through interference with Mg uptake, transport, and metabolism and (5) damage to lipids and proteins through reaction of Mn with sulfhydryl groups and activation of lipoxygenases.

The importance of photo-oxidation in Mn phytotoxicity is supported by several lines of evidence. High light intensity potentiates Mn toxicity in common bean and sugar maple. In common bean, excess Mn induces antioxidant enzyme systems, especially in high light, and a Mn-tolerant genotype is better able to maintain reduced ascorbate pools than a Mn-sensitive genotype under Mn stress.

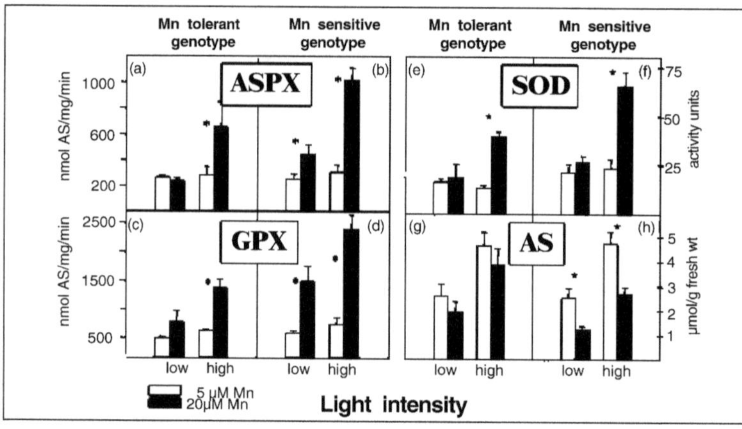


Interaction of Mn and light intensity on antioxidant systems in common bean. In leaves with excess Mn, high light intensity upregulates the activity of ascorbate peroxidase (a, b), guaiacol peroxidase (c, d), cytosolic superoxide dismutase (e, f), and decreases leaf ascorbate (g, h), especially in a Mn-sensitive genotype. Asterisks indicate significant differences between Mn-stressed and control plants. These responses, as well as those of other systems not shown here, indicate that excess Mn creates photo-oxidative stress, and that genetic variation in antioxidant metabolism (in this case, maintenance of reduced ascorbate pools), may be related to Mn tolerance.

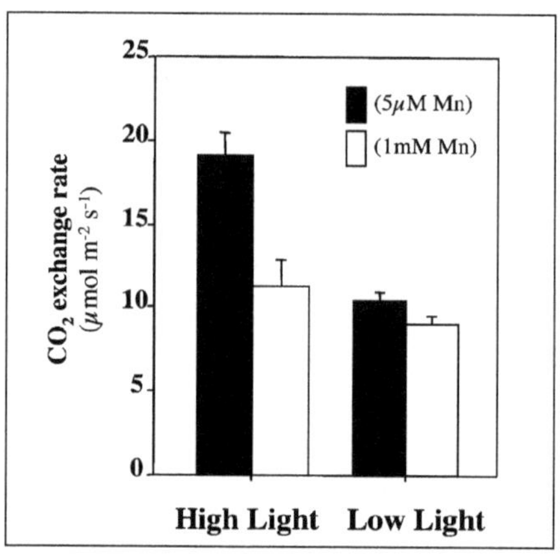


Interactive effects between excess Mn and light intensity (1500 mmol $m^{-2}$ $s^{-1}$ vs. 300 mmol $m^{-2}$ $s^{-1}$) on photosynthesis of sugar maple seedlings.

Rice (Oryza sativa L.) chloroplasts produce more active oxygen species under Mn stress, and tobacco (Nicotiana tabacum L.) cell lines selected for Mn tolerance have elevated guaiacol peroxidase activity. The role of oxidative stress in Mn phytotoxicity explains prior reports of synergism between Mn toxicity and light intensity, and Mn toxicity and ozone exposure.

The role of photo-oxidative stress in Mn toxicity has important implications for the effects of global change on ecosystems prone to Mn stress, such as the Eastern Forest of North America. A number of environmental variables that are changing as a result of human influences interact with oxidative stress, including increased UV radiation, temperature stress, changing light levels, drought stress, and ozone exposure. These changes are occurring concurrently with increasing Mn bioavailability in forest soils caused by acid deposition and base leaching. Because of the strong interaction of Mn toxicity with the metabolism of other metals, atmospheric factors that affect the uptake of these nutrients may have important interactions with Mn phytotoxicity. Increasing atmospheric $CO_2$ should decrease leaf transpiration, and thereby reduce the uptake of Ca and Mg by mass flow to the roots, in addition to changing Ca and Mg allocation to expanding leaves by changing xylem fluxes. These effects may exacerbate Mn inhibition of leaf metabolism of Ca and Mg. Through photo-oxidation, tree responses to Mn toxicity may be greatly magnified in response to other variables such as increasing light intensity. Substantial interspecific differences in tolerance to Mn toxicity and other nutrient stresses could lead to changes in species composition, and possibly changes in ecosystem function. Therefore, Mn toxicity and related photo-oxidative processes may play a central role in integrating ecosystem responses to a changing climate.

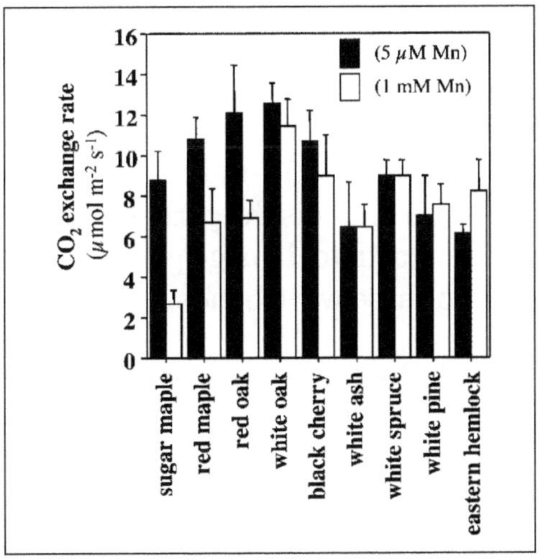


Effect of toxic levels of manganese on photosynthesis of tree seedlings grown in sand culture. Sugar maple, red maple and red oak—three keystone tree species of the Eastern Forest were the most sensitive.

## OXIDATIVE STRESS

Oxidative stress is a complex chemical and physiological phenomenon that accompanies virtually all biotic and abiotic stresses in higher plants and develops as a result of overproduction and accumulation of reactive oxygen species (ROS). Recent data have clarified the 'origins' of oxidative stress in plants, and show that apart from classical chloroplast, mitochondrial and peroxisome sources, ROS are synthesized by NADPH oxidases and peroxidases. ROS damage all major plant cell bio-polymers, resulting in their dysfunction. They activate plasma membrane $Ca^{2+}$-permeable and $K^{+}$-permeable cation channels as well as annexins, catalyzing $Ca^{2+}$ signaling events, $K^+$ leakage and triggering programed cell death. Downstream ROS-$Ca^{2+}$-regulated signaling cascades probably include regulatory systems with one (ion channels and transcription factors), two ($Ca^{2+}$-activated NADPH oxidases and calmodulin) or multiple components ($Ca^{2+}$-dependent protein kinases and mitogen-activated protein kinases). Intracellular and extracellular antioxidants form sophisticated networks, protecting against oxidation and 'shaping' stress signaling. Research into plant oxidative stress has shown great potential for developing stress-tolerant crops. This can be achieved through the use of directed evolution techniques to prevent protein oxidation, bioengineering of antioxidant activities as well as modification of ROS sensing mechanisms.

## ULTRAVIOLET RADIATION STRESS

The Earth receives the Sun's energy in the form of electromagnetic radiation. Humans are able to see only a small portion of this radiation, i.e., visible light (400- 700 nm). The vast majority of sunlight is beyond human perception. Ultraviolet rays are just outside the realm of visible light and are characterized by wavelengths shorter than the last visible light ray, violet, hence "ultraviolet." UV-radiation that reaches Earth's surface is in the wavelengths between 280-400 nm. It is the part of the electromagnetic spectrum that can reach a high enough level on Earth to damage plants ecosystem. Ultraviolet radiation has been arbitrarily divided into three wave bands termed UV-A, UV-B and UV-C.

UV-C (100 to 280 nm) can be found in the artificial sources such as mercury arc lamps and germicidal lamps. UV-C is completely absorbed by the ozone layer and oxygen.

UV-B (280 to 315 nm) is the most intense usually between the hours of 10:00 am to 3:00 pm when the sunlight is brightest. It is much intense in the summer months accounting for about 60-70% of a trees yearly UV-B dose. UV-B does not penetrate the glasses. UV-B is mostly absorbed by the ozone layer, but some does reach the Earth's surface.

UV-A (315 to 400 nm) radiation is more constant than UV-B without any variation during the day and throughout the year. UV-A is neither filtered by glass nor absorbed by the ozone layer.

Shorter the wavelength, more energetic is the radiation causing damages to the trees. UV-C can do great damage but fortunately poses no risk to life on Earth. It is completely filtered out by oxygen and ozone in the stratosphere. Ozone also plays a vital role in filtering UV-B, the ultraviolet radiation that is the greatest threat to life on Earth. However, even a fully functioning ozone layer does not absorb all the UV-B rays. UV-A surpasses the stratosphere virtually unfiltered, but compared to the shorter wavelengths, UV-A causes little harm. If the ozone layer thins further, the increased UV penetration will have severe consequences for trees life and other life forms as well. Substantial reductions of upto 50% in the ozone column have been observed over Antarctica. In the tropics, due to small solar zenith angle and thin ozone layer, plants encounter much higher level of UV-radiation than at higher latitude. Increased UV radiation could inhibit tree growth and development, photosynthesis, stunting plant life and agriculture. On a clear day, the level of UV radiation reaching the ground depends on how direct the sunrays are. Rays hitting the Earth directly, naturally, have more energy than less direct rays. They move through less energy-filtering atmosphere than lower angled rays.

On Long Island in ideal conditions, the greatest intensity of UV radiation will reach the area in summer solstice during the solar noon when the Sun is at its highest point in the sky for the entire year. This is the moment in time when the sunrays are most direct. Before and after this date and time, the sun hits the area at lower angles, less directly, decreasing the radiation. In general, the closer one moves towards the equator, more direct the sunrays will be. That area will receive, on average, high levels of UV radiation for the year. Atmosphere is rarely completely clear, and even it is clear, it contains various materials that will limit radiation. Dust particles and air molecules floating in the air absorb radiation and also reflect it scattering and deflecting it back into space. The higher the elevation the more unfiltered radiation that area will receive. A high elevation simply lies above many of the dust particles and air molecules preventing the radiation's passage. In addition, if an area is high enough, it will also perch above another factor limiting radiation, clouds.

## Factors Affecting UV Radiation

The level of UV radiation reaching on the Earth surface depends upon a number of factors and UV-radiation at the earth's surface is modified by temporal, geographical and meteorological factors. Following factors influence the risk of UV radiation exposure and its consequences on tree health and life:

- Ozone Layer: The ozone layer absorbs most of the sun's UV rays, but the amount of absorption varies depending on the time of the year and other natural phenomena. The absorption of UV radiation has decreased, as the ozone layer has become thinner due to the release of ozone-depleting substances that have been widely used in industry viz., CFCs, (green house gases) etc.
- Time of the Day: When the sun is at its highest in the sky around noon, sun rays have the least distance to travel through the atmosphere and UV-B levels are at the highest. During early morning and late afternoon, sunrays pass through the atmosphere at an angle and travel longer distance and their intensity is reduced.
- Time of the Year: The sun angle varies with the season and causes the intensity of UV rays to change. During summer months, UV intensity tends to be the highest.
- Cloud Cover: Clouds play a big role in allowing the amount of UV radiation reaching the Earth's surface. On a cloudy day, depending on the shape and

thickness of the clouds, they can absorb and reflect 35-80 % of the Sun's radiant energy and along with the other effects prevents all but a negligible amount of radiation from reaching the ground. Even with heavy cloud cover the scattered UV component of sunlight is seldom less than 10% of that under clear sky.

- Latitude: At equator sunrays are the strongest, where it is most directly overhead and UV rays have to travel the least distance through the atmosphere. Ozone layer is usually thinner in the tropics compared to the mid and high-latitudes, so there is less ozone to absorb the UV radiation, and hence more UV radiation passes through the atmosphere. At higher latitudes, the Sun is lower in the sky, so UV rays must travel a long distance through ozone-rich portions of the atmosphere and, in turn, trees grown at these latitudes are less exposed to UV radiation.
- Altitude: At higher altitude the UV intensity increases because there are fewer atmospheres to absorb the damaging rays. In general, each 1 km increase in altitude increases the UV flux by about 6%. The trees grown at higher altitudes have higher risk of overexposure to UV radiation.
- Other Factors: Additional factors limiting UV radiation include haze, smog, surface reflection and other air pollution. All these factors are taken into account in a scale, called the UV index, measuring the amount of UV radiation reaching the ground and its relative danger to plant life.

## Impact of Increased UV Radiation on Tree

Forests occupy about 31 percent of the Earth's land area and make up over 90 percent of the Earth's biomass. Forests account for two-third of the carbon that is "fixed" or withdrawn from the atmosphere. Forests thus play a major role in how much carbon is free in the atmosphere, which in turn affects the magnitude of the "greenhouse effect." A little information exists on the effects of UV-B radiation on forest tree species. Tropical forests, though representing nearly one-half of the global productivity and much of the total tree species diversity, have received very little attention with respect to the ozone reduction problem. Although little, or no, ozone reduction has thus far occurred in the tropics, only a small decrease of ozone at these latitudes would result in a very sizeable increase of UV-B radiation since solar UV-B radiation is already very intense in these regions. Study has shown that excluding existing solar UV-B radiation with filters can result in increased growth of four broadleaf species. Otherwise, the effects of UV-B radiation on tropical tree species have not received much attention. The perception of increased UV-B radiation has been diverted by global climate change; the problem has not gone away. In fact, even if ozone-depleting emissions were halted immediately, the detrimental gases already in the stratosphere break down slowly. Studies on agricultural species have shown that about 60 percent are at least moderately sensitive to high levels of UV-B radiation.

One of the primary interests of a tree physiologist is what effect UV-B has on ribulose 1,5-bisphosphate carboxylase oxygenase (RuBisCO), the most abundant protein on Earth. It is believed that increased carbon dioxide and global warming will offer a buffer against UV-B damage to a certain extent. Increased carbon dioxide can enhance plant growth, but when combined with the negative effect of UV-B radiation on photosynthetic enzymes in C3 and C4 plants, the result is far from certain. Physiologists are worried about whether the increased UV-B radiation will change carbon allocations within trees. They may have to put much of their photosynthetic products into protective mechanisms at the expense of growth. Growth processes uses photosynthetic assimilate and are controlled by content and activity of hormones and inhibitors. In most cases the treatment of plants with UV leads to the shift of hormone inhibitor balance towards inhibitor predominance. There could be more severe direct effects also, but considering the role of trees in regulating atmospheric carbon, even small effects could in turn have large effects on climate change.

UV-B affects trees by modifying both biological and chemical environment. The most important of these are the nucleic acids and to a much lesser extent the proteins and other molecules. Damage may occur in a number of ways, including the direct destruction of the genetic material DNA, deactivation of enzymes, disruption of membranes and other cell structures and the generation of highly reactive chemical agents known as free radicals. Elevated UV exposure can cause temporary or irreversible damage to photosynthetic apparatus (including the bleaching of the pigments), loss of electron transport efficiency to processes of cell division and growth regulation, and the composition and replication of genetic material. Consequences include a reduction in growth yield, changes in levels and effects of plant hormones and alteration of periods of dormancy, flowering, etc. Biological repair mechanisms exist, mutations may remain as errors in the repair processes. In addition, the repair mechanisms themselves may be deactivated by high UV doses. The interaction of all these processes can lead to a variety of adverse effects on plants. Many effects are sub-lethal, may interact with other factors and therefore, very difficult to attribute to UV-enhancement specifically.

## Molecular and Cellular Constitution

DNA readily absorbs UV-B radiation, which commonly changes the shape of the molecule in one of several ways. Changes in the DNA molecule often mean that protein-building enzymes cannot read the DNA code at that point on the molecule. As a result, distorted proteins can be made, or cells can die. However, living cells are "smart." Over millions of years of evolving in the presence of UV-B radiation, cells have developed the ability to repair DNA. Enzymes such as, BER (base excision repair), NER (nucleotide excision repair) and photolyase arrive at the damage site, remove the damaged section of DNA, and repair it with the proper components (based on information elsewhere on the DNA molecule). This makes DNA somewhat resilient to damage by UV-B.

Exposure of plants to UV-B radiation results in changes in expression of a large number of genes. Before UV-B radiation or light of other wavelengths can give rise to a cellular response, it has to be perceived by some kind of receptor, and the information transduced via a signalling pathway to the target molecules, be it proteins in the cytoplasm or the genetic material in the nucleus. Elevated solar UV-B doses increase the frequency of somatic homologous DNA rearrangements in plants. Increases in recombination are accompanied by a strong induction of photolyase and Rad51 gene expression. These genes are putatively involved in major DNA repair pathways, photoreactivation and recombination repair. In plants that are deficient in photoreactivating ultraviolet-induced cyclobutane pyrimidine dimers, recombination under elevated UV-B regimes greatly exceeds wild-type levels. The results showed that homologous recombination repair pathways might be involved in eliminating UV-B-induced DNA lesions in plants. Increased solar UV-B radiation is forecasted for the early 21st century may affect genome stability in plants.

Scots pine seedlings were exposed in a phytotron to simulated outdoor climates with ambient or twice-ambient levels of ozone (26-96 and 52-192 nL $L^{-1}$, respectively) combined with zero or low ambient levels of UV-B. Needles and cotyledons were evaluated at eight dates up to 44 days of treatment. Cinnamylalcohol dehydrogenase (CAD) and stilbene synthase (STS) were induced by ozone. The mRNAs of these stress-related enzymes were induced only under twice-ambient ozone concentrations. Ozone-induced transient STS transcript levels reached their maximal values between day 1 and day 5, and were more pronounced in the presence of UV-B. CAD mRNA content and CAD enzyme activity increased continuously under enhanced ozone concentrations. Additional UV-B resulted in suppressed transcript level and enzymic activity of CAD. The results indicated that at the level of gene expression, there is an interaction between responses to UV-B and ozone.

Suesslin et al. have observed that ultraviolet-B radiation triggers the activation of protection mechanisms and various photomorphogenic responses. One of the Arabidopsis mutants, uli3, shows a reduced sensitivity of UV-B induced gene expression events. The ULI3 gene encodes an 80 kDa protein with potential domains for heme and diacylglycerol-binding. In transiently transfected protoplasts, the ULI3: GFP fusion protein is localized in the cytoplasm but also adjacent to membranes. In etiolated seedlings irradiated by UV-B, ULI3 mRNA expression is strongly up regulated. This is caused by elevated transcription as demonstrated using stable transformants where a GUS-reporter is driven by the ULI3-promoter. ULI3 is preferentially expressed in the outer cell layers in leaves, stems and flowers, but not in roots. There is evidence that ULI3 represents a specific component involved in UV-B-mediated signal transduction in higher plants.

The perception of low levels of UV-B probably occurs via a UV-B photoreceptor, followed by several different signalling pathways. These pathways include second messengers such as calcium, kinases and the catalytic formation of reactive oxygen species. On other hand, high levels of UV-B probably cause cellular damage and oxidative stress, thus activating a general stress signal transduction pathway, which leads to a response similar to that, which occurs after pathogen attack, and other stresses. Some of the genes identified so far as being regulated by UV-B encode proteins involved in the biosynthesis of protective pigments, DNA repair and antioxidative enzymes, photosynthetic genes, cell cycle genes, and stress genes induced by other types of stimuli i.e., pathogenesis-related proteins and senescenceinduced genes.

UV-B radiation damages a variety of cellular components like the membrane of leaf cells, particularly the structure of chloroplast membrane. Under UV-B radiation, the plasma membrane in Schefflera octophylla was contracted and corrugated, and separated from the cell wall. The chloroplast in S. octophylla and Pterospermum heterophyllum was dilated and even broken and the thylakoid dilated and adhered. In Archidendron clypearia, the mitochondria appeared to be cavitated. Starch grains in chloroplast in A. clypearia and P. heterophyllum were increased, and the black particles in chloroplast in S. octophylla and Sapium discolor were also increased, and appeared on the mitochondrial membrane of S. octophylla.

A tree's ability to resist the effects of UV-radiation is a product of its ability to reduce exposure (through the optimisation of growth pattern and the production of UV absorbing pigments) and to repair or replace damaged molecules. The expression of these UV resistance mechanisms is sensitive to the quantity and quality of ambient light. The significance of DNA repair as a UV resistance mechanism was investigated by comparing the effects of solar UV on the growth of a variety of repair-proficient and deficient Arabidopsis lines grown under natural light, focusing on the effects of UV on the growth of well-established plants. UV had an inhibitory effect on the growth of wild-type plants, and this effect was enhanced in repair-defective lines. Plants defective in the cyclobutane dimer photolyase are most sensitive to the effects of UV-B on plant height and rosette diameter growth, indicating that the enzyme plays an important role in UV resistance. A mutant defective in nucleotide excision repair (NER), a non-specific repair pathway, also displayed a significant, though less sensitivity to solar UV-B, suggesting that either the transcription-coupled repair of UV-induced dimers, or the repair of some other UV-induced lesion was also important for optimal growth.

## Biochemical Constitution

Increased UV-B radiation exposure to the plants leads to biochemical changes in the leaves, such as secondary compounds, which are having UV-absorbing properties like condensed tannins, several flavonoids, and some hydrocinnamic acid contents are increased. The solar UV-B radiation plays an important role in the formation of secondary metabolites in the leaves of birch seedlings scot pine seedlings and leaves of woody tree species. Seasonal variation in accumulation of UV-B absorbing compounds noticed is more correlated to the developmental processes than to the seasonal fluctuations of the UV-B radiation.

Secondary compounds, such as lignin, are important as structural materials. These are related to phenolic compounds and may change in composition with elevated UV-B radiation. If the ratio of lignin to cellulose in plant tissues changes, it can alter the rate of decomposition. The seedlings of Acer saccharum and Acer platanoids exposed to UV-A radiation accumulated higher concentrations of anthocyanins and other flavonoids. The roots synthesized more UV-absorbing compounds than shoots; coumarin, aesculetin, scoparone and various phenolic acids were identified in the root extracts.

The prolonged exposure of enhanced UV-B radiation to seven-year-old Norway spruce seedlings caused significant decrease in chlorophyll a and b content. This decrease was mainly caused by the decrease in chlorophyll a content (up to 21%). This decrease in photosynthetic pigment caused reduced photosynthetic gas exchange. The antenna size also reduced due to long-term exposure to UV-B radiation. Concentration of foliar phenolics increased in populus when exposed to enhanced UV-B radiation. Exposure of Leucadendron laureolum seeds to UV-B radiation resulted in decreased seed germination, which was associated with decreased foliar chlorophyll a and carotenoid levels in the seedlings.

One important conifer species, loblolly pine has been shown to be particularly sensitive to UV-B radiation. Loblolly pine responded to UV-B radiation under a full solar spectrum by increasing needle flavonoid content. Two-year-old tissue had higher foliar flavonoids content under enhanced UV-B radiation in Quercus rubra, Pinus ponderosa and Psedotsuga menziesii under field condition.

Plant parts in addition to leaves, accumulate specific compounds in response to UV-radiation exposure. Concentration of sucrose, raffinose and glucose in the bark of silver birch were higher in UV treated saplings than in saplings grown in ambient radiations. Saplings grown in the elevated UV-A + UV-B radiations had significantly increased concentration of certain UV-B absorbing phenolics, such as salidroside, 3,4-dihydroxypropiohenone-3-glucoside, (+)-catechin and (-)-epicatechin compared with saplings grown in ambient radiation.

## Growth and Development

The timing of life phases of plants is a combination of response to environmental factors and the genetic constitution of the plant. UV-B exposure can alter the timing of events such as flowering, entering and breaking of dormancy, and even senescence is important not only to the individual plant, but to also how plants interact with other plants. For example, a shift in the timing of flowering can mean that a plant species may not have sufficient insect pollinators available at the new time of flowering because either the insects are not present or other plant species are attracting these pollinators. Such changes could also conceivably be important in agricultural systems, but intervention with management options may make these changes less important. Increased UV-B may advance or delay (depending on species) the time of flowering in tree species. There is little work at present on flowering responses and virtually nothing on other potential effects of UV-B on life phase timing of tree species.

Exposure of Populus clones to near ambient UV-B radiation caused shift in allocation of carbon from leaf development to other pools probably to the UV absorbing compounds (flavonoids derivatives). Such reallocation curtails leaf development and reduces photosynthetic capacity compared to subambient levels of UV-B radiation and affect growth over long duration. In Populus deltroids trees when grown under enhanced UV-B radiation, height, diameter and biomass reduced significantly, but shoot to root ratio increased.

Five-year-old European beech trees were exposed to increased UV-B radiation and ambient UV-B radiations. The increased level of UV-B had adverse effect on net photosynthesis (Pn), stomatal conductance (gs), chlorophyll fluorescence (Fv/Fm) and accelerated senescence measured as yellowing of leaves. Significant reduction in plant height, leaf area and shoot dry weight was observed when Salvia splendens plants were grown under enhanced UV-B radiation level. Reduction in leaf area ratio and simultaneous increase in specific leaf area was also observed due to enhanced UV-B radiation, which indicated that UV-treatment might have induced significant accumulation of photosynthates in leaf blades.

Leaf anatomical changes are important in determining the degree of injury sustained by plants exposed to natural and enhanced levels of UV-B radiation. The degree to which leaf anatomy changes due to increased UV levels is poorly understood in most of the tree species. Four tree species were exposed to enhanced UV-levels and examined for change in leaf anatomy. Populus tricocarpa, an intermediate broad leaf species, showed significantly thicker palishade parenchyma in recently matured leaves and increased leaf area. Quarcus rubra, a semi determinate broad leaf species, exhibited significantly thicker palisade parenchyma as compared to control. Psuedotisuya menziesii a coniferous species with flattened needles, and Pinus ponderosa, an evergreen coniferous species with a complete hypodermis, showed no significant changes in the leaf area or specific leaf weight under enhanced UV-B radiation.

Loblolly pine is the leading commercial species in the south-eastern United States and accounts for a large portion of pulp producing capacity. Loblolly pine is one of the most susceptible species to UV-B radiation. When loblolly pine seedlings were exposed to enhanced UV-B radiation a reduction of 40 and 16%, respectively was observed in biomass and height. In Douglaus fir seedlings, increased lateral branching was noticed. Studies have demonstrated deleterious effects of UV-B radiations on tree growth and physiology. Shift in dry matter allocation towards roots contributed to lower shoot/root and leaf area ratio in UV-B treated silver birch seedlings. However, UV-B radiation inhibited the growth of main roots and stems, decreased number of lateral roots and leaf number as well as leaf area in Psychotria rubra, Schefflera octophylla, Archidendron clypearia, Pterospermum heterophyllum, Sapium discolor and Albizia lebbeck seedlings.

## Mineral Nutrition

Uptake and translocation of mineral nutrients within the plant can be affected by elevated UV-B radiation. Consequently, the mineral nutrient status of plants can affect plant responsiveness to UV-B radiation. Nitrogen concentration in plant tissues can increase under elevated UV-B, which is linked with reduced insect herbivory.

Growth and pigment accumulation of plants exposed to UV-radiation are dependent upon the availability of mineral nutrients. Young scot pine seedlings grown at different nutrient levels were exposed to UV-A and UV-B radiation. The growth and the accumulation of photosynthetic pigments were positively affected by the availability of nutrients, while the accumulations of condensed tannins and its precursor, (+)-catechin decreased significantly under high nutrient level. Nutrient level did not markedly affect the total accumulation of flavonoid. Silver birch seedlings when exposed to enhanced UV-B radiations, a marked reduction in nitrogen allocation to leaves and increased allocation of phosphorus to roots was noticed.

Tosserams et al. reported that an increase in carbon accumulation in nutrient stressed plants led to a reduction in UV-B induced damages because of increased foliar UV-B absorbance by enhanced accumulation of phenolic compounds and leaf thickening. Total biomass production of plants at low nutrient supply was 50% lower as compared to plants grown at high nutrient supply, while net photosynthesis decreased only by 12%. Increased levels of UV-B reduced biomass production under non-limiting nutrition conditions only and there was no effect on the plants grown under limiting nutrient condition. This response was associated with accumulation of carbohydrates under nutrient limited nutrition condition.

## Photosynthesis

Effects of UV radiation, particularly those of the biologically more active UV-B radiation on trees and specifically on photosynthesis have been extensively investigated. Studies with artificial UV-B sources have shown that UVB radiation may strongly affect PSII, whereas PSI appears to be relatively insensitive. Reductions in photosynthetic rate and capacity have been described for a number of tree species. Experiments have shown that as little as 16 per cent decrease in current ozone level could result in 40 per cent reduction in the productivity of Loblolly pine. Similar effects are reported for jack-pine, spruce, sycamore, birch and ash; however, the effects are normally more severe in conifers than in deciduous species. Absorption of light in excess of photosynthetic utilization by green plant leaves may lead to a reduction in the potential efficiency of PSII, which persists in low light or darkness and is regarded as the major cause of photoinhibition of photosynthesis. It has been shown for many plant species that photoinhibition of photosynthesis does occur under natural conditions. In studies with tropical plants, substantial reductions in the potential efficiency of PSII, indicated by a decline in the "dark-adapted" ratio of variable to maximum chlorophyll (Chl) a fluorescence (Fv/Fm) have been observed upon direct exposure to the solar beam of outer crown leaves of mature forest trees and of plants situated in treefall gaps within the tropical forest.

UV-B photons inactivate PSII via non–interacting mechanism, which affects different target sites. When UV-B radiation is accompanied by low intensity visible light, which provides protection against photodamage results in synergistically enhanced protein repair capacity and protects PS-II damage. However, this ameliorating effect becomes insignificant at high light intensities characteristic of direct sunlight.

UV-B radiation inhibited $CO_2$ assimilation when seedlings of Anacardium exculsum, Virola surinamensis and Colophyllum longifalium were exposed to direct sunlight as compared to UV-B filtered sunlight. The rate of photosynthetic uptake by the leaves was affected which occurred due to inhibition of potential PSII efficiency. The exposure to UV-B radiation also decreased the efficiency of P 700 photooxidation. However, electron transport, RuBisCO activity and enzymes from the Calvin cycle were highly sensitive to UV-B radiation along with low temperature stress in Norway spruce trees, which affected the $CO_2$ fixation level adversely.

The UV-A component of sunlight is also a significant damaging factor for photosynthesis, which targets PSII complex. UV-A radiation results in rapid inhibition of oxygen evolution accompanied by the loss of signal from the S2 state of the water-oxidizing complex. UV-A irradiance also affects the relaxation of flash induced variable chlorophyll fluorescence. The primary damage site of UV-A irradiance is the catalytic manganese cluster of the water-oxidizing complex, where electron transfer to Tyr-Zo and P680+ is inhibited. The damaging mechanism of UVA radiation is very similar to that induced by the shorter wavelength UV-B radiation, but different from photosynthetically active longer wavelength radiation (400-700 nm). Primary damaging site of UV-B is the water-oxidizing complex. It also induces degradation of D1 and D2 protein subunits of PSII reaction center.

Sun facing leaves from the outer crown of the mature trees are less susceptible to UV-stress in comparison to leaves in shade. Photoinhibition of PSII occurred when shade leaves of tropical trees were exposed to UV-light. Substantially less inhibition of PSII occurred when UV-B radiation was excluded and was further reduced when both UV-B and UV-A were excluded. The effects of long-term UV-B radiation exposure on the growth and morphology of woody perennials, the gas exchange and photosynthesis of five common deciduous tree species (Fraxinus excelsior, Betula pendula, Tilia cordata, Quercus robur and Acer pseudoplatanus) were measured. All five tree species had been exposed to UV-B radiation for 5 years in the field at an enhancement level equivalent to 18% ozone depletion. Measurements made during the fifth year of UV-B irradiation recorded reductions in light-saturated photosynthesis, transpiration and water use efficiencies. These changes were accompanied by marked reductions in individual leaf area, stomatal density and conductance and carboxylation efficiency. There were no significant changes in the maximum variable fluorescence ratio, the quantum requirement of oxygen evolution, or light-saturated $O_2$ production. Analysis of the response of net carbon assimilation to changing intercellular $CO_2$ concentration (A/ci response) demonstrated no significant change in stomatal limitation. Reductions in photosynthesis were consistent with decreased carboxylation efficiency.

Krause et al. studied the effect of UV-A and UV-B on the leaves of tropica trees. The ambient UV-B and UV-A radiation, observed upon exposure to full, direct sunlight contributed to the reversible decline in potential PSII efficiency. Sensitivity of PSII to natural UV-B light depended upon the acclimation status and developmental stages of the leaves and tended to decline with increased levels of vacuolar UV-absorbing compounds. In shade-grown tree seedlings periodically exposed to full sunlight, protein damage was indicated by the strong effects of ambient PAR and UV light on Fv/Fm and fluorescence induction kinetics, and by the extremely slow recovery. Leaves of the crown of mature trees that are acclimated to fullsunlight may achieve full protection against adverse effects of UV on PSII. But at certain stages of development, or depending on the prior light environment (incomplete acclimation), ambient UV light may cause temporary losses in PSII efficiency. Although tropical plants are capable of effectively protecting their leaves against UV stress, an anthropogenic rise in ambient UV-B radiation in the tropics may markedly affect their PSII efficiency.

## UV Radiation and other Stress Factors

Plants and other organisms in nature are seldom affected by only a single stress factor, such as UV-B radiation. Instead, they typically respond to several factors acting in concert, such as water stress, increased atmospheric $CO_2$, mineral nutrient availability, heavy metals, tropospheric air pollutants and temperature. Therefore, it is important to keep in mind that the effectiveness of UV-B radiation can be greatly increased or decreased by such factors. Visible radiation is an important ameliorating factor and thus, natural levels as possible should be applied in laboratory experiments for attaining realistic results.

## Water Stress

Among the most common of factors in nature is water stress. Interactions between UV-B radiation and water status of plants also occur. Elevated UV-B radiation in field experiments tended to alleviate drought symptoms in two Mediterranean pine species. UV-B radiation acts as an environmental signal to induce tolerance to high light drought stress. Douglas-fir seedlings grown with 6 kJm–2 day–1 of biologically effective UV-B, which is ambient for the inter mountain regions of Idaho, USA, have higher light saturated carbon assimilation rates than seedlings grown without UV-B radiation. UV-B radiation along with drought stress (withholding water for 15 days) had 50% high light saturated carbon assimilation rate and 40% high seedlings water potential.

## Increased $CO_2$ Concentration

Increases of atmospheric $CO_2$ are a certain element of global climate change and atmospheric $CO_2$ concentration will likely double by the middle of the next century. Many experiments with elevated $CO_2$ employ a twice-ambient $CO_2$ concentration as a treatment condition. Such a doubling often results in more pronounced plant responses than are evident in many elevated UV-B radiation lamp experiments designed to simulate up to 20% ozone column reduction under field conditions. However, responses to $CO_2$ are small in semi-natural ecosystems where nutrient or water availability may strongly constrain plant growth. For example, Gwynn-Jones et al. showed that growth responses to elevated $CO_2$ and enhanced UV-B (both alone and in combination) were small during the first three years of experimentation in a sub-arctic heath. Also, most ecosystem-level effects of elevated $CO_2$ are mediated through changes in plant tissues. When studied independently, plant growth responses to changes in UV-B radiation and atmospheric $CO_2$ concentration generally are thought to be in opposite directions. Usually, in most experiments employing both elevated $CO_2$ and UV-B radiation, these factors do not yield interactions, with some exceptions. Elevated $CO_2$ sometimes appears to provide some protection against elevated UV-B radiation for some species; yet, elevated UV-B radiation can limit the ability of some species to take advantage of elevated $CO_2$ in photosynthesis. Allocation of biomass in plants can also change in a complicated fashion with the combination of $CO_2$ and UV-B radiation treatments. Increased temperature is also a predicted element of global climate change. In a study combining two levels of UV-B radiation with two levels of $CO_2$ and two temperatures, the results indicated that either elevated $CO_2$ or somewhat high temperature had similar effects in reducing the growth-inhibiting effects of elevated UV-B radiation.

In trees, the allocation of carbon to secondary metabolites has been shown to be determined by both the availability of resources (e.g., $CO_2$ concentration) and by specific stress factors (e.g., UV radiation). In combination, $CO_2$ and UV-B radiation may differentially affect tree growth and morphogenic parameters, and elevated $CO_2$ may ameliorate the effects of UV-B radiation. The effects of increased atmospheric $CO_2$ concentration and UV-B radiation on growth and the accumulation of different types of secondary metabolites were studied in silver birch (Betula pendula). Seedlings were exposed to 350 and 700 μmol mol-1 of $CO_2$ in a greenhouse. At both $CO_2$ levels, additional UV-B was either present (8.16 kJ $m^{-2}$ $day^{-1}$ of biologically effective UV-B irradiance) or absent. The time course of accumulation of individual secondary compounds and the shifts in allocation of carbon between biomass and the secondary metabolites (phenolic acids, flavonoids, condensed tannins) were studied during a month-long exposure. The activities of enzymes (L-phenylalanine ammonia-lyase [PAL], EC 4.3.1.5; peroxidase, EC 1.11.1.7; polyphenol oxidase, EC 1.10.3.1) were also determined for leaves. UV-B radiation significantly increased biomass, PAL activity, and the accumulation of phenolic acids and flavonoids in seedlings. Elevated $CO_2$ concentration increased the activities of all the enzymes and the accumulation of condensed tannins in leaves, especially with UV-B radiation. Because the observed UV-B induction of flavonoids was small under a high $CO_2$ concentration, it was suggested that the excess of carbon in the atmosphere might moderate the effect of UV-B by increasing the metabolic activity of leaves (high enzyme activities) and changing the allocation of internal carbon between different primary and secondary metabolites in the plant.

## Oxidative Stress

The nature and origin of the reactive oxygen species (ROS) are involved in the early part of UV-B-induced signalling pathway. The increase in PR-1 transcript and decrease in Lhcb transcript in response to UV-B is mediated through pathways involving hydrogen peroxide derived from superoxide ($O^{\cdot -2}$). The origins of the ROS were also shown to be distinct and to involve NADPH oxidase and peroxidases. The up-regulation of Chs by UV-B was not affected by ROS scavengers, but was reduced by inhibitors of nitric oxide synthase (NOS) or NO scavengers. This suggests that UV-B exposure leads to the generation of ROS, from multiple sources, and NO through increased NOS activity, giving rise to parallel signalling pathways mediating responses of specific genes to UV-B radiation.

Four-year-old Scotts pine (Pinus sylvestris) and Norway spruce (Picea abies) seedlings potted in natural soil were exposed in phytochambers to fluctuating $O_3$ concentrations and UV-B radiation. After one growing season, current-year needles were collected and analysed for activities of antioxidative enzymes (superoxide dismutase, SOD, EC 1.15.1.1; catalase, CAT, EC 1.11.1.6; guaiacol peroxidase, POD, EC 1.11.1.7) and soluble antioxidants (ascorbate and glutathione). CAT, POD, ascorbate and glutathione, but not SOD, were increased in needles of both species in response to twice-ambient $O_3$ levels. UV-B radiation in the presence of ambient $O_3$ caused an increase in total SOD activity in spruce but had no effect on antioxidants in pine. Twice ambient $O_3$ levels together with low UV-B radiation counteracted the $O_3$ induced increases in ascorbate and CAT in pine but not in spruce. Under these conditions, spruce needles showed the highest antioxidative protection and revealed no indication of lipid peroxidation. Pine needles exposed to UV-B and elevated $O_3$ levels showed elevated lipid peroxidations and a 5-fold increase in dehydroascorbate, suggesting that this species was less protected and suffered higher oxidative stress then spruce.

Active oxygen formation (AOS) increased in the embryos of Fucus spiralis exposed to enhanced UV-B. Two photoinhibition responses were recognized: (i) a rapid decline of the PSII yield due to the violoxanthin-zeaxanthin cycle (photoprotection), and (ii) a slower second-phase decline, correlated with active oxygen production. The electron transport rate increased with embryo age, and was correlated with AOS production. Enhanced AOS production was associated with slow recovery of the PSII yield, with increased effective UV dose.

## Insects

The extent to which plant tissues are consumed by insects or the degree to which pathogens attack plants is regulated by several properties of the plant host tissues. Experiments show that present-day solar UV-B radiation can substantially reduce insect herbivory of agricultural and native plant foliage. Field studies involving supplementation of solar UV-B radiation with lamp systems indicated a substantial reduction in populations of a herbivorous insect on a heathland plant. The reasons for these changes are not always clear, but they may be mediated through changes in plant secondary chemistry or alterations in plant nitrogen or sugar content. Studies involving UV lamps indicated decreased herbivory by a moth caterpillar under elevated UV-B radiation and this was attributed to increases of host pea plant tissue nitrogen content. Mulberry plants previously irradiated with UV from lamps suffered less herbivory by silkworms (Bombyx mori) and the low consumption was attributed to low sucrose content of the foliage. In laboratory, studies McCloud and Berenbaum have shown that UV-B radiation can increase furanocoumarin content of plant tissue, which, in turn, results in slow development of certain insect larvae during early life stages of the larvae. Thus, the foregoing would suggest that insect herbivory might always be decreased by UV-B radiation. The results of most of these studies indicate that the effects on insect herbivory are all due to biochemical changes in the host plant tissues. However, there are some indications that some insects may respond directly to solar UV-B radiation.

## Plant Pathogen

Plant fungal and viral diseases react in several different ways to UV-B radiation in various experiments, conducted primarily in laboratory and greenhouse conditions. In four of ten studies, UV-B was found to counteract disease severity and in the other six studies, it promoted disease development. The direction of the UV-B radiation effect on disease severity can also vary with the variety of the host. It is not clear in many of these experiments whether the changes in disease severity are due simply to changes caused by UV-B radiation in the host plant, or direct UV-B radiation effects on the fungal or viral pathogens was involved. Plants exposed to UV-B radiation prior to infection were more susceptible to subsequent infection by fungal pathogens, while UV-B irradiation after infection had no effect on disease severity. Such an experiment suggests that the effect of UV-B radiation is mediated through changes in the host plant tissues. There is also evidence from solar UV-B exclusion studies that incidence of fungal disease increases when UV-B is removed.

Pedunculate oak (Quarcus robur) saplings were exposed to ambient levels of UV-A and UV-B radiation and controlled solar radiation. The effect on leaf fungi was examined weekly over a four month-period using two techniques; the spore fall and leaf impression methods, which differentiated between those fungi occurring on the upper (adaxial) and lower (abaxial) surfaces of the leaves. The abundance of Aureobasidium pullulans and Sprobololymes roseus, two leaf yeasts which had adaxial: abaxial ratio of <1 under ambient levels of UV-B radiations was negatively correlated with increasing ambient level of UV-B radiations and reduced significantly on adaxial leaf surfaces by the supplemental UV-B radiation. Alternaria and Monographelle nivalis showed consistent responses on adaxial surfaces to UV-A radiation applied over control + treatment array. These results suggest that the current levels of shortwave radiation already influence the distribution of fungi on leaf surfaces and further increase will have more detrimental effect on leaf fungi distribution.

These changes in insect herbivory and disease severity caused by alterations of solar UV-B can be sizeable; they can operate in different directions and have very important implications for both agricultural and non-agricultural ecosystems. They may be much more important than known influences of UV-B radiation on plant production based on realistic field studies. Even roots of plants whose shoots are exposed to elevated UV-B radiation can be affected as indicated by root interactions with microorganisms. For example, the nature of microorganism assemblage associated with roots of sugar maple trees (Acer saccharum) was altered by exposure of the tree shoots to elevated UV-B radiation. This was obviously a systemic effect of UV-B expressed in the roots of the host plant.

## Interaction between Species

The primary plant productivity in forests and grasslands may not be greatly affected by ozone reduction even if the growth of some plants is diminished. The plant species differ greatly in growth responsively to UV-B. It is anticipated that a productivity reduction of one species will probably lead to increased productivity of another relatively more UV-tolerant species. This is likely because more resources will be available to the tolerant species. Thus, the overall productivity of the system may well remain about the same, while species composition may change. However, a change in the balance of species could have far-reaching consequences for the character of many ecosystems.

Another mechanism whereby the competitive balance of plant species can be changed by increased UV-B is through changes in plant form. Even if plant production per se is not affected by increased UV-B, changes in plant form can result in changes in which species can more effectively compete for sunlight. A quantitative analysis of competition for sunlight in the mixed stands with and without supplemental UV-B showed that subtle changes in plant form of the two species were sufficient to change the balance of competition for sunlight that is necessary for photosynthesis. Therefore, one species can achieve some advantage over the other because one captures more sunlight for photosynthesis. In these experiments, the wheat benefited from increased UV-B and weed suffered. However, in other mixtures of crop and weeds, the situation might be reversed. Also, other changes in plant form, such as greater allocation of biomass to roots might change competitive effectiveness of individual species for soil moisture and nutrients. In grasslands and forests that are not managed intensively, similar changes in species composition may be experienced. Ecosystem-level experiments with nonagricultural systems are only beginning. Early reports of one experiment in a subarctic heath ecosystem suggest that species composition changes may result from UV-B supplementation.

## Impact of UV-B A at different Latitudes and Altitudes

UV-B irradiance is naturally very low in high-latitude ecosystems, such as tundra and sub arctic areas. There is experimental evidence that the plants in such systems react to increased UV-B associated with realistic levels of ozone depletion. Some plant species exhibit growth inhibitions and others do not, thus, eventually altered community composition is expected. Long-term observations of species composition are being pursued in high-latitude subarctic systems in Sweden, a high arctic site on Spitzbergen Island and in southernmost Argentina. In the latter system, attenuating the naturally occurring solar UV-B radiation increased insect herbivory, decreased plant tissue nitrogen concentrations and increased populations of some microfauna. The subarctic studies in Sweden have been underway for years and show various effects including decreased litter decomposition, increased fruit formation and greatly increased insect herbivory.

Although terrestrial ecosystems at high latitudes are not highly productive for grazing, timber production, etc., the influence of ozone reduction on these systems is important for several reasons. Carbon sequestration is generally quite high in these ecosystems, including the extensive peat formations that are also being studied in the Swedish subarctic and southern Argentinean systems. Compared with other locations, these ecosystems are under the greatest ozone depletion, especially in the Southern Hemisphere, and they also experience the greatest warming as the global greenhouse effect intensifies. Thus, they are sensitive indicators of several features of climate change.

## References


- Feng Z, Guo A, Feng Z. 2003.Amelioration of chilling stress by triadimefon in cucumber seedlings // Plant Growth Regulation Journal.– vol. 39, No. 3, p. 277–283
- Chaves, M. M.; Flexas, J.; Pinheiro, C. (2009-02-01). "Photosynthesis under drought and salt stress: regulation mechanisms from whole plant to cell". Annals of Botany. 103 (4): 551–560. doi:10.1093/aob/mcn125. ISSN 0305-7364. PMC 2707345. PMID 18662937
- Zauralov OA, Lukatkin AS. 1997. Aftereffect of low temperatures on the respiration of heat-loving plants // Russian Journal of Plant Physiology. –vol. 44, No. 5, p. 640–644
- Lambers, Hans; Chapin II, F. Stuart; Pons, Thijs L. (2008). Plant physiological Ecology. New York, NY: Springer. p. 349. ISBN 978-0-387-78340-6
- Salinity-stress-and-salt-tolerance, abiotic-stress-in-plants-mechanisms-and-adaptations: intechopen.com, Retrieved 13 April, 2020

# Factors Affecting Stress Tolerance

CHAPTER 4

There are diverse factors affecting stress tolerance that are studied under the domain of biotic and abiotic stress in plants. It includes transcription factors, role of fungal endophytes, soil bacteria and endophytic microbes against biotic and abiotic stress tolerance. This chapter discusses the aspects related to various factors affecting stress tolerance in detail.

## ROLE OF PLANT TRANSCRIPTION FACTORS IN ABIOTIC STRESS TOLERANCE

Plants are constantly exposed to a wide range of environmental stresses such as drought, high salt, heat and extremes of temperature. Growth constraints due to these abiotic stresses result in reduced productivity and significant crop losses globally. Drought and salinity affect more than 10% of arable land, which results in more than 50% decline in the average yields of important crops worldwide. Tolerance or susceptibility to these stresses is also a very intricate event as stress may affect multiple stages of plant development and often several stresses concurrently affect the plants. Therefore, the basic mechanisms of abiotic stress tolerance and adaptation have been the area of comprehensive research.

Plants counter adverse environmental conditions in a complex, integrated way depending on the timing and length that allows them to respond and adapt to the existing constraints present at a given time. Plant stress tolerance involves changes at whole-plant, tissue, cellular, physiological and molecular levels. Exhibition of a distinct or a combination of intrinsic changes ascertains the capacity of a plant to sustain itself under unfavorable environmental conditions. This comprises a range of physiological and biochemical adjustments in plants including leaf wilting, leaf area reduction, leaf abscission, root growth stimulation, alterations in relative water content (RWC), electrolytic leakage (EL), production of reactive oxygen species (ROS) and accumulation of free radicals which disturb cellular homeostasis ensuing lipid peroxidation, membrane damage, and inactivation of enzymes thus influencing cell viability. Other than these, abscissa acid (ABA), a plant stress hormone, induces leaf stomata closure, thus reducing transpirational water loss and photosynthetic rate which improves the water-use efficiency (WUE) of the plant. Molecular responses to abiotic stress on the other hand include perception, signal transduction, gene expression and ultimately metabolic changes in the plant thus providing stress tolerance.

Several genes are activated in response to abiotic stresses at the transcriptional level, and their products are contemplated to provide stress tolerance by the production of vital metabolic proteins and also in regulating the downstream genes. Transcript profiling can be a significant tool for the characterization of stress-responsive genes. Extensive transcriptome analyses have divulged that these gene products can largely be classified into two groups. First group comprises of genes that encode for proteins that defend the cells from the effects of water-deficit. These genes mainly include those that regulate the accumulation of compatible solutes (enzymes for osmolyte biosynthesis like proline, betaine, sugars, etc.); passive and active transport systems across membranes (water channel proteins and membrane transporters); and protection and stabilization of cell structures from damage by ROS (the detoxification enzymes such as glutathione S-transferase, catalase, superoxide dismutase, ascorbate peroxidase, etc.); fatty acid metabolism enzymes, proteinase inhibitors, ferritin and lipid-transfer proteins; and other proteins for the protection of macromolecules (LEA protein, osmotin, chaperons, etc.). Another group of genes stimulated by abiotic stresses includes regulatory proteins that further regulate the stress signal transduction and alter gene expression and hence possibly function in stress response. They comprise several transcription factors (TFs) emphasizing the role of various transcriptional regulatory mechanisms in the stress signal transduction pathways; protein kinases (MAP kinase, CDP kinase, receptor protein kinase, etc.); protein phosphatases and proteinases implicated in the regulation of stress signaling and gene expression.

## Role of Transcription Factors in Abiotic Stress Responses

Transcription factors (TFs) are proteins that act together with other transcriptional regulators, including chromatin remodeling/modifying proteins, to employ or obstruct RNA polymerases to the DNA template. Plant genomes assign approximately 7% of their coding sequence to TFs, which proves the complexity of transcriptional regulation. The TFs interact with cis-elements in the promoter regions of several stress-related genes and thus up-regulate the expression of many downstream genes resulting in imparting abiotic stress tolerance. In Arabidopsis thaliana genome about 1500 TFs are described which are considered to be involved in stress responsive gene expression. Transcriptome data in Arabidopsis and in numerous other plants suggest that there are several pathways that independently respond to environmental stresses (in both ABA dependent and independent manner), suggesting that stress tolerance or susceptibility is controlled at the transcriptional level by an extremely intricate gene regulatory network.

The phytohormone ABA is the central regulator of abiotic stress particularly drought resistance in plants, and coordinates a complex gene regulatory network enabling plants to cope with decreased water availability. ABA-dependent signaling systems have been illustrated as pathways that mediate stress adaptation by induction of atleast two separate regulons (a group of genes controlled by a certain TF):

- The AREB/ABF (ABA-responsive element-binding protein/ ABA-binding factor) regulon.
- The MYC (myelocytomatosis oncogene)/MYB (myeloblastosis oncogene) regulon.

While ABA-independent regulons are:

- The CBF/DREB regulon.
- The NAC (NAM, ATAF and CUC) and ZF-HD (zinc-finger homeodomain) regulon.

However in addition, several studies have identified the existence of both ABA-dependent and –independent pathways of stress response that function through AP2/EREBP (ERF) family members. In addition to these well-known regulons, a large number of other TFs are also involved in abiotic stress responses, thereby playing a crucial role in imparting stress endurance to plants.

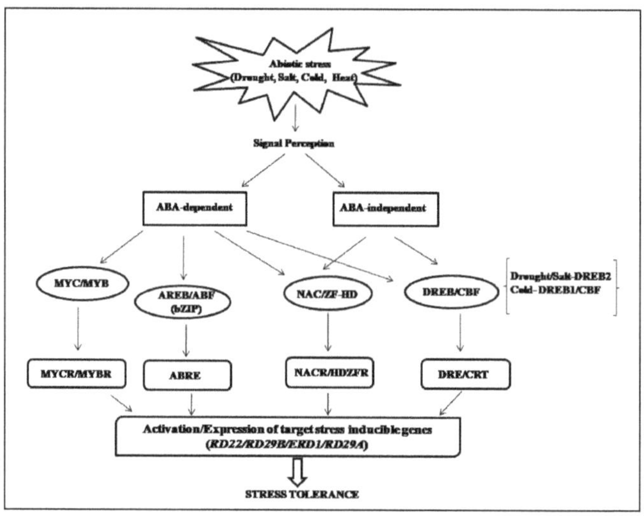


A schematic representation of transcriptional regulatory networks of cis-acting elements and transcription factors involved in abiotic-stress-responses. Transcription factors are shown in ellipses; cis-acting elements are shown in boxes; and target stress inducible genes are shown in long rectangular box at the bottom.

Although these different stress responsive TFs usually function independently, it is undoubtedly possible that some level of cross-talk exists between them.

This chapter focuses on these TFs and their role in regulating abiotic stress responses in plants as well as their utility in engineering stress tolerance for crop improvement programs.

## The AREB/ABF Regulon

A conserved cis-element named as ABA-responsive element (ABRE; PyACGTGG/TC) was identified from the promoters of ABA-inducible genes. Subsequently it was revealed that ABA-responsive gene expression needs multiple ABREs or the combination of an ABRE with a coupling element (CE) as a functional promoter. For example, ABRE and coupling elements, including coupling element 1 (CE1) and coupling element 3 (CE3), constitute an ABA-responsive complex in the regulation of wheat HVA1 and HVA22 genes. For the expression of RD29B in seeds and vegetative tissues of Arabidopsis, two ABRE cis-acting elements are required.

The AREB or ABFs are bZIP (basic leucine zipper) TFs that bind to the ABRE motif and activate ABA-dependent gene expression were first isolated in a yeast one-hybrid screening. It was reported that in the ABA-deficient aba2 and ABA-insensitive abi1 mutants, the AREB/ABF proteins have less activity while they show an enhanced activity in the ABA hypersensitive era1 mutant of Arabidopsis suggesting that these TFs require an ABA-mediated signal for their activation. The reason possibly may be an ABA-dependent phosphorylation of the AREB/ABF proteins. The 75 AtbZIPs have been divided into 11 groups, and the ABFs/AREBs are classified to group A which usually act in ABA signaling during seed maturation or stress conditions. Several studies have suggested that ABFs function in different stress response pathways; i.e. ABF1 in cold; ABF2 in salt, drought, heat and glucose; ABF3 in salt; ABF4 in cold, salt, and drought signaling pathways. AREB/ABFs are phosphorylated by ABA-responsive 42-kDa kinases which suggest that ABA-dependent phosphorylation may be involved in activation of AREB subfamily proteins. These kinases (SnRK2-type) such as OST1/SRK2E in Arabidopsis phosphorylate Ser/Thr residues of R-X-X-S/T sites in the conserved regions of AREB1.

AREB/ABF genes are mostly redundant in tissue-specific expression either in vegetative tissues or seeds. AREB1/ABF2, AREB2/ABF4, and ABF3 were mainly expressed in vegetative tissues, whereas ABI5 and EEL were expressed during seed maturation and germination. Rice homolog TRAB1 and barley homolog HvABI5 activated ABA-responsive gene expression in seeds. Expression of OsABI5 was stimulated by ABA and high salinity, but was down-regulated by drought and cold stress in seedlings, and its overexpression also improved salinity tolerance in rice. ZmbZIP17 was up-regulated by drought, heat, ABA and NaCl stress in maize seedlings.

Overexpression of ABF3 and ABF4 resulted in reduced transpiration and improved drought tolerance. AREB1/ABF2 was found to be a crucial component of glucose signaling, and its over-expression improved drought stress tolerance. Overexpressing OsbZIP23, a member of AREB/ABF subfamily can also significantly improve drought and high salinity resistance of transgenic rice at the reproductive stage. Enhanced tolerance to drought and heat was also observed in 35S-OsAREB1 transgenic Arabidopsis plants. The over-expression of the constitutively active form of AREB1 in transgenic Arabidopsis plants showed ABA hypersensitivity and enhanced drought tolerance, and LEA-class genes and ABA and dehydration-stress-inducible regulatory genes such as linker histone H1 and AAA ATPase were upregulated. Over-expressing SRK2C caused hypersensivity to ABA, improved drought tolerance and lowered transpiration rate. Overexpression of AtbZIP60 led to improved salt tolerance.

## The MYC/MYB Regulon

The MYC/MYB families of proteins are universally found in both plants and animals and known to have varied functions. Both MYC/MYB TFs participate in the ABA-dependent pathway of stress signaling for the upregulation of the abiotic stress responsive genes. The first MYB gene identified was the v-MYB gene of avian myeloblastosis virus (AMV). The first plant MYB gene, C1, was identified in Zea mays. It encodes a c-MYB-like TF that is involved in anthocyanin biosynthesis. Wide existence of MYB genes indicates that these are very ancient evolutionarily. A MYB domain is usually composed of one to three imperfect repeats, each with about 52 amino acid residues that adopt a helix-turn-helix conformation intercalating in the major groove of the DNA. Plant MYB proteins are categorized into three major groups:

- R2R3-MYB having two adjacent repeats.
- R1R2R3-MYB having three adjacent repeats.
- MYB-related proteins, usually containing a single MYB repeat.

The R2R3 family contains the largest number of MYB genes. have reported that there are 198 and 183 MYB genes in the Arabidopsis and rice genomes, respectively.

MYB TFs play important roles in many physiological processes under normal or unfavorable growth conditions and also in secondary metabolism, cell morphogenesis, meristem formation and floral and seed development, cell cycle control, defense and stress responses, and hormone signaling. MYC and MYB TFs accumulate only after ABA accumulation. AtMYB4 (At1g22640), AtMYB6 (At4g09460), AtMYB7 (At2g16720), AtMYB44 (At5g67300), AtMYB73 (At4g37260), AtMYB77 (At3g50060), and AtMYBCDC5 (At1g09770) were found to be constitutively expressed in all organs and during all stress treatments. AtMYB2 and AtMYC2 function cooperatively as transcriptional activators in the dehydration and ABA-inducible rd22 expression. According to, AtMYB102 integrates dehydration, osmotic, or salinity stress, ABA application, and wound-signaling pathways. AtMYB60 and AtMYB61are involved in light-induced opening of stomata and dark-induced closure of stomata, respectively. AtMYB44, AtMYB73, and AtMYB77 are activated by wounding, white-light, cold stress, and salt stress. AtMYB44 and AtMYB77 expression is reduced in fus3 (fusca3), lec1 (leafy cotyledon1), and abi3 (ABA-insensitive3) mutants that are defective in development of dormancy and drought tolerance during late embryogenesis and seed maturation. AtMYB44 TF confers abiotic stress tolerance through enhancing stomatal closure in an ABA-independent manner. Recent studies have shown that AtMYB15 expression is detectable in both vegetative and reproductive organs and is up-regulated by cold and salt stresses. AtMYB15 has been found to negatively regulate freezing tolerance in Arabidopsis with its ability to repress the expression levels of CBF genes. AtMyb41 from Arabidopsis is transcriptionally regulated in response to salinity, drought, cold, and ABA. Identified 156 GmMYB genes of which the expression of 43 genes changed on treatment with ABA, salt, drought and cold stress.

Overexpression of MYB15 results in improved drought and salt tolerance in Arabidopsis. Increased expression levels of AtMYB2, AtMYC2 or both enhance ABA sensitivity and improve osmotic tolerance. Overexpression of 35S:AtMYC2 and 35S:AtMYB2 and 35S:AtMYC2+AtMYB2 in Arabidopsis induced ABA responsive stress genes and showed an ABA-hypersensitive phenotype with increased osmotic stress tolerance. Transgenic plants overexpressing AtMyb41 showed dwarf phenotype due to alterations of cell expansion and cuticle integrity and enhanced drought sensitivity. Overexpression of AtMyb75 and AtMyb90 led to increased anthocyanin levels, while Met-derived glucosinolate content of Arabidopsis increased with overexpression of AtMyb28. In contrast, OsMYB3R-2 transgenic plants showed enhanced tolerance to freezing, drought and salt stress and decreased sensitivity to ABA. Different level of tolerance was imparted by overexpression of OsMYB4 depending on the nature of the host plants. Arabidopsis transgenic plants overexpressing OsMYB4 showed increased chilling and freezing tolerance with a dwarf phenotype, the tomato transgenic showed higher tolerance to drought stress, whereas increased drought and cold tolerance was observed in the apple transgenic. Overexpression of a StMYB1R-1 transgene in potato plants improved plant tolerance to drought stress while having no significant effects on other agricultural traits.

## The CBF/DREB Regulon

The dehydration responsive element binding proteins (DREBs) are important AP2/ERF plant TFs that induce a set of abiotic stress-related genes, thus imparting stress tolerance to plants. These play an important role in the ABA-independent pathways that activates stress responsive genes. The first isolated cDNAs encoding DRE binding proteins, CBF1 (CRT binding factor1), DREB1A and DREB2A were identified through yeast one-hybrid screening from Arabidopsis. Since then, many DREBs have been isolated from various plants. These proteins specifically bind to and activate the expression of genes regulated by the DRE sequence (5'-TACCGACAT-3') and were first identified in the promoter of the drought-responsive gene rd29A. DREB1 and DREB2 are two main subgroups of DREB subfamily, involved in two different signal transduction pathways under cold and dehydration respectively. DREB1B/CBF1, DREB1A/CBF3 and DREB1C/CBF2 genes are positioned in consonance on chromosome 4 of Arabidopsis. Arabidopsis also contains major DREB2 proteins namely, DREB2A and DREB2B. DREB1/DREB2-homologous genes have also been identified in various cereals and millet crops.

The DREB TFs contain an extremely conserved AP2/ERF DNA-binding domain throughout plant kingdom. The domain consists of a three-stranded β-sheet and one α-helix running almost parallel to it that contacts DNA via Arg and Trp residues located in the β-sheet. Two conserved functional amino acids (valine and glutamic acid) at $14^{th}$ and $19^{th}$ residues respectively, exist in the DNA binding domain, which are crucial sites for the binding of DREBs and DRE core sequences. An alkaline N-terminal amino acid region that serve as a nuclear localization signal (NLS) and a conserved Ser/Thr-rich region responsible for phosphorylation near the AP2/ERF DNA binding domain are also mostly present. The proteins contain an acidic C-terminal region which might be functional in trans-activation activity.

The activation of these transcripts is organ-specific and comparative to the extent of the stress given. When exposed to salt stress, AhDREB1 was highly expressed in roots but less significantly in stems and leaves. It was observed that OsDREB1F was constitutively expressed throughout the plant with highest expression in panicles and callus than in the other tissues. AtDREB2A accumulated in roots, stems and leaves under control conditions. DREB2C expressed in mature embryo and the cotyledons of germinating seedlings reported that sunflower HaDREB2 expresses in all vegetative tissues. Chrysanthemum DvDREB2A was expressed in all organs under normal conditions. SiDREB2, a DREB2 gene accumulated in leaves, roots, young and mature spikelets of foxtail millet indicating its function in developmental pathways also.

AtDREB1 was induced within 10 min at 4°C. The transcript of CBF genes was detectable after 30 min at 4°C with highest accumulation at 1 h. HvDREB1 gene in barley leaves significantly accumulated on salt, drought, and low-temperature treatments. OsDREB1A and Os-DREB1B were induced early (within 40 min) after cold exposure but not on ABA treatment. OsDREB1A was induced within 5 h of salinity stress whereas OsDREB1C showed constitutive expression. PNDREB1 strongly responded to low temperature and dehydration. However, hot pepper Ca-DREBLP1 was quickly activated by dehydration, high salinity and mechanical wounding but not at all by cold stress. The expression of Arabidopsis DREB2A and its homolog DREB2B were stimulated by dehydration and high salinity, but not by cold and ABA. Similarly, ABA, mannitol and cold treatments had minimal effect on DREB2C expression. A detailed study of all five rice OsDREB2s showed that OsDREB2A expressed to the highest levels under the control condition, and its expression was increased to some extent by high temperature, drought and high salinity, but not by low temperature treatments. Expression of OsDREB2B was markedly increased after 20 min of high and 24 h of low temperature stress. While the transcript levels of OsDREB2C, OsDREB2E and OsABI4 were low under the control condition and were transiently induced by the abiotic stresses.

Wheat TaDREB1and WDREB2, maize ZmDREB2A, and pearl millet PgDREB2 are responsive to cold stress while foxtail millet SiDREB2 was not. Expression of chickpea CAP2 was induced by dehydration, NaCl, ABA and auxin treatments but not by low temperature, salicylic acid and jasmonic acid. The transcript expression of Salicornia brachiata SbDREB2A was stimulated by NaCl, drought and heat stress.

Transgenic Arabidopsis plants over-expressing DREB1B/CBF1 or DREB1A/CBF3 show strong tolerance to freezing, drought, and high salinity stresses implying that DREBs/CBFs affect multiple genes. DREB1A/CBF3 overexpressing transgenics accumulated proline and various sugars under non-stress conditions. Transgenic Arabidopsis and rice plants overexpressing OsDREB1A too displayed tolerance to low temperatures, high salinity and drought. The rd29A:DREB1A/CBF3 wheat transgeneic showed improved drought stress tolerance. Likewise, the constitutively overexpressing CBF3/DREB1A and ABF3 transgenic rice showed better drought and salinity tolerance without any growth inhibition or phenotypic anomalies. The overexpression of AhDREB1 accumulated putative downstream target genes and also conferred improved survival rate to transgenic tobacco plants under salt stress as compared to the wild-type plants. The over-expression of OsDREB1F greatly enhanced tolerance of plants to high salinity, drought, and low-temperature both in rice and Arabidopsis, thus playing a significant role in plant stress signal transduction. Microarray analysis of transgenic Arabidopsis plants suggested that over-expression of DREB2A-CA induced drought-, salt-responsive and heat-shock (HS)-related genes. These transgenic plants also exhibited enhanced thermotolerance which was significantly decreased in DREB2A knockout plants. Overexpression of DREB2C was also found to activate the expression of many HS responsive genes. Transgenic Arabidopsis plants overexpressing maize ZmDREB2A were dwarf and also displayed improved drought and heat stress tolerance. Transgenic Arabidopsis plants overexpressing OsDREB2B showed enhanced expression of DREB2A target genes and improved drought and heat-shock stress tolerance. Transgenic tobacco plants overexpressing PgDREB2A showed better tolerance to both hyperionic and hyperosmotic stresses. Transgenic tobacco plants overexpressing CAP2 showed improved growth and development, and tolerance to dehydration and salt stress. While its expression in yeast (Saccharomyces cerevisae) enhanced heat tolerance, with increased expression of heat shock factor 1 (Hsf1) and its target yeast heat shock protein 104 (Hsp 104) suggesting strong evolutionary conservation of the stress response mechanisms. In another remarkable study it was described that the recombinant E. coli cells expressing SbDREB2A exhibited better growth in basal LB medium as well as if supplemented with NaCl, PEG and mannitol.

These studies indicate that the DREB proteins are important TFs in regulating abiotic stress-related genes and play a critical role in imparting stress endurance to plants.

## The NAC (NAM, ATAF and CUC) and ZF-HD (Zinc-finger Homeodomain) Regulon

The NAC family of plant-specific TFs is one of the largest in the plant genome, with 106 and 149 members in Arabidopsis and rice, respectively. NAC family TFs contains a highly conserved N-terminal DNA-binding domain and a diversified C-terminal domain. NAC was derived from the names of the first three described TFs containing NAC domain, namely NAM (no apical meristem), ATAF1-2 and CUC2 (cup-shaped cotyledon). The cis-element of NAC TF [NAC recognized sequence (NACRS)] was also identified in Arabidopsis.

Numerous studies have examined the involvement of several types of NAC TFs in plant developmental programs, and disease resistance. A few NAC genes were found to be involved in response to various environmental stresses also such as ANAC019, ANAC055, and ANAC072 from Arabidopsis, and BnNAC from Brassica. SNAC1 is activated mainly in guard cells under dehydration. AtNAP and its homologs play an important role in leaf senescence in Arabidopsis. ERD1 promoter analysis showed that TFs belonging to the NAC family and ZF-HD are important for the activation of the ERD1 (early responsive to dehydration stress 1) gene. XND1 is expressed in xylem and associated with stress, ABA response and leaf senescence in Arabidopsis. In soybean 101 NAC domain containing proteins, identified as functionally non-redundant were involved in response to abiotic stresses and in cell death events whereas GmNAC2, GmNAC3 and GmNAC4 were strongly induced by osmotic stress. Soybean NACs GmNAC3 andGmNAC4 were also induced by ABA, JA and salinity but differed in their response to cold. GmNAC1, GmNAC5 and GmNAC6 transiently expressed in tobacco leaves, resulting in cell death and enhanced expression of senescence markers. Flavonoid biosynthesis is regulated by ANAC078 under high-light. A rice NAC gene, ONAC045 was induced by drought, high salt, low temperature, and ABA treatment in leaves and roots. The transcription level of CaNAC1 could be elevated by exogenous SA, ET, and MeJA treatment. A novel wheat NAC TF, TaNAC4 was found to be induced in response to cold, salt, wounding, ABA, ethylene and MeJA, suggesting a significant cross-talk between abiotic and biotic stress conditions reported that a salt-inducible NTL8 (membrane associated NAC) regulates gibberellic acid-mediated salt signaling in seed germination. Very recently a membrane associated NAC TF from foxtail millet was found to be up-regulated in drought, salinity, ethephone and MeJA treatments.

Several target genes of the ANAC019, ANAC055, and ANAC072 transcriptional activators were identified in the Arabidopsis transgenic plants using cDNA microarray. These transgenic plants also exhibited improved drought tolerance. The SNAC1-overexpressing transgenic rice seedlings showed significantly higher survival rate than wild type under drought treatment and significantly enhanced salinity tolerance as well. A rice R2R3-MYB gene (UGS5) containing putative NACRS in the promoter region was also induced in the SNAC1-overexpressing plants. Many abiotic and biotic stress responsive genes were up-regulated in the OsNAC6 transgenic plants, and the transgenics were tolerant to dehydration, high salt stresses. ONAC045 overexpressing rice plants showed enhanced tolerance to drought and salt treatments. XND1overexpressing showed severe stunting, premature death, and repression of TE differentiation. Hence NAC TFs play an indispensable role in physiological adaptation for successful plant propagation under abiotic stress conditions.

## Other TFs in Abiotic Stress Response and Tolerance

There are a number of TFs which are involved in abiotic stress responses other than the TFs belonging to the well known regulons described above. A new class of homeodomain TF known as HIGHER EXPRESSION OF OSMOTICALLY RESPONSIVE GENE 9 (HOS 9) and a R2R3-type MYB protein HOS 10 have been identified recently which are found to be associated with cold stress. hos9 and hos 10 mutants show freezing hypersensitivity but at the same time enhance expression of RD29A and other cold responsive genes without changes in the CBF/DREB1 regulon implicating their role as negative regulators of cold stress-responsive genes. Another homeodomain TF, HDG11 which codes for HD-START TF plays a significant role in drought tolerance by enhancing the water homeostasis of the plants.

HARDY (HRD), an AP2-EREBP IIIc TF gene is expressed in inflorescence tissue to protect it from desiccation. Rice plants overexpressing HRD exhibited drought and salinity tolerance as well as improved WUE. ERFs (ethylene responsive factors) also belong to the AP2-EREBP TF and have been found to be involved in growth, development, metabolic regulation and biotic and abiotic stress responses. Transgenic tobacco plants expressing SodERF3 exhibited extremely improved drought and salt tolerance reported that transgenic tobacco plants overexpressing soybean GmERF3 exhibited tolerance not only to high salinity and drought stresses but also to various pathogens, suggesting its crucial role in both abiotic and biotic stresses.

WRKYs are another important class of plant TFs which have shown to possess multiple functions in plants including abiotic stress responses. OsWRKY45 in rice was up-regulated by dehydration, cold, heat and salt. Arabidopsis overexpressing OsWRKY45 also showed improved drought tolerance. They have suggested that OsWRKY45 may be involved in ABA synthesis that induces a signaling cascade resulting in lowered transpiration and enhanced tolerance to drought. Overexpression of the OsWRKY89 in rice led to growth inhibition at early stages of plant development, but showed increased tolerance to UV irradiation and fungal infection. GmWRKY13, GmWRKY21 and GmWRKY54 were found to be differentially expressed under abiotic stresses. Transgenic Arabidopsis plants overexpressing GmWRKY21were tolerant to cold stress, whereas GmWRKY54 conferred salt and drought tolerance, possibly through the regulation of DREB2A and STZ/Zat10. However, transgenic plants over-expressing GmWRKY13 showed increased sensitivity to salt and mannitol stress, decreased sensitivity to ABA, and an increase in lateral roots. Reported that down-regulation of NbWRKY; an abiotic stress related WRKY TF, by virus-induced gene silencing produced chlorosis and senescing phenotype in tobacco plants.

Zinc finger proteins (ZFPs) are one of the important TFs found abundantly in plants and animals. They contain sequence motifs in which cysteines and histidines coordinate zinc atom(s) forming local peptide structures required for their specific functions. Cys2/His2 (C2H2)-type ZFPs containing the EAR transcriptional repressor domain, play a key role in regulating the defense responses of plants to biotic and abiotic stress conditions. Over-expression of Alfin1, a novel member of the ZFP family confers salt tolerance to the transgenic Alfalfa plants. The constitutive over-expression of soybean SCOF-1 induced cold-regulated (COR) gene expression and transgenic Arabidopsis and tobacco plants. ZPT2-3, a C2H2-type Petunia ZFP, when constitutively over-expressed in petunia, resulted in dehydration tolerance of transgenic plants. OSISAP1 from rice was inducible by cold, desiccation, salt, submergence, heavy metals and wounding, and its overexpression in tobacco exhibited cold, dehydration and salt tolerance at the seed germination/seedling stages. Constitutive expression of Zat12 in Arabidopsis resulted in the increased expression of oxidative and light stress responsive genes. Transgenic Arabidopsis plants constitutively expressing the Zat7 exhibited suppressed growth and were more tolerant to salinity stress. CaZF, a C2H2 ZFP provided salinity-tolerance in transgenic tobacco. Interestingly, heterologous expression of CaZF provided osmotolerance in S. cerevisiae through Hog1p and calcineurin dependent as well as independent pathways.

# ABIOTIC STRESS TOLERANCE IN PLANTS BY ENDOPHYTIC MICROBES

Endophytes are microorganisms residing inside healthy plant tissues without causing any apparent harm to the host. Wilson proposed that endophytes are fungi or bacteria, which, for all or part of their life cycle, invade the tissues of living plants and cause unapparent and asymptomatic infections entirely within plant tissues but cause no symptoms of disease. Based on the nature of pathogenicity, endophytes may be of three types:

- Pathogens of another host that are non-pathogenic in their endophytic relationship.
- Non-pathogenic microbes.
- Pathogens that have been rendered nonpathogenic but still capable of colonization by selection methods or genetic alteration.

The term endophyte was first used by de Bary in 1866 to denote all those microbes residing inside the living healthy plants. While few years before de Bary, in 1846 Leveille recognized the fungi in wheat leaves and called them as endophytic fungi.

## Sources of Endophytic Colonization

The source of endophytic colonization ranges from transmission via seeds and vegetative planting material to entrance from the surrounding environment such as the rhizosphere and phyllosphere. Dong et al., 1994 have demonstrated bacterial transmission via vegetative planting material for Acetobacterdiazotrophicusfor two successive cuttings of sugarcane while occurrence of endophytic bacteria in healthy seeds has been reported for some plants. High similarity between rhizosphere and phyllospheremicrobes found internally in the root with those found in the rhizosphere or phyllospherehas been indicated to be the main sites of endophytic colonization.

## Entrance and Establishment of Endophytes within Plant Tissues

Endophytes follow the same pattern as the pathogenic microorganisms to enter into the plants. The expression of defence related molecules in host is found less at the time of entry of endophytes than pathogens.With the exception of some endophytes transmitted by seed or vegetative propagation material, successful endophytic colonization goes through several important stages including host finding, recognition, colonization of the plant surface and entrance into the internal plant tissue.Endophytic microbes enter plants through natural openings suchas hydathodes, stomata and lenticels,wounds caused by abrasion with soil particles, pathogen damage, and formation of lateral roots, micropores and abiotic mechanical damage. As a consequence of Rhizoctonia damage ofPhaseolus vulgaristhetwo endophytes Enterobacter asburiaeJM22 and Pseudomonas fluorescens89B-27 extensively colonized fungal necrotic tissue.In order to penetrate the cuticle and cell walls, microbes produce cell wall degrading enzymes such as cutinase, pectinase, cellulase, hemicellulase, protease, and lignin-peroxidases to enter into the plant. The expression of a proteinase was identified and purified during infection of endophytic Acremonium typhinum to the grass Poa ampla. This proteinase was abundant within fungal membrane vesicles and in the plant and fungal cell walls at the time of infection while absent in fungus culture. In case of root endophytic fungus Piriformospora indica, it interferes with the host cell death program to form a mutualistic symbiosis with plants. The already established plant endophytes may transmit from one generation to another generation via plant seeds. The rhizospheric bacteria may enter and establish as root endophytes through emergence of lateral roots or root hair cells, primary and lateral root cracks, and diverse tissue wounds occurring as a result of plant growth.The genotype of host and fungus is an important factor to determine the establishment of symbiotic relation. A PCR based genetic analysis of Venturia ditricha, a fungal endophyte of birch trees has explained that the infection frequency by a particular endophyte genotype is influenced by the host genotype.

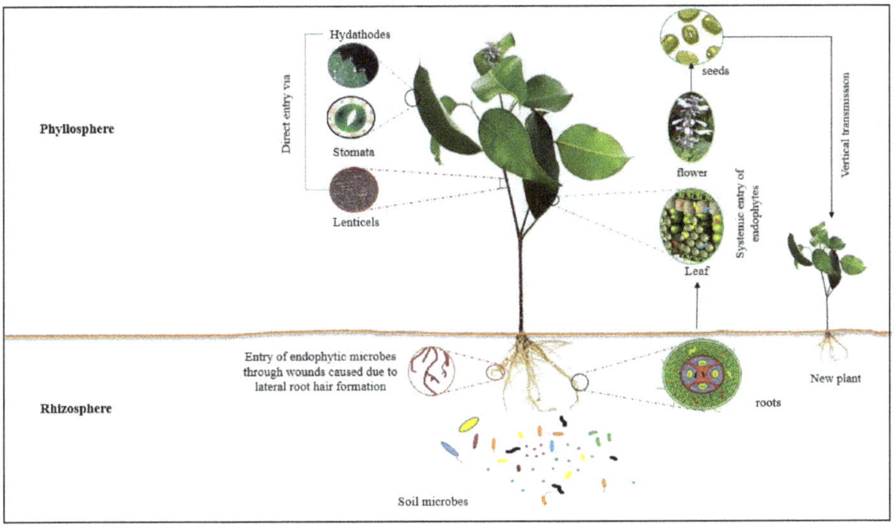


Bacterial entry into plants.

## Endophytes as Stress Controller in Plants

The symbiotic interactions between plant and endophyte may result in several outcomes as defined by fitness benefits by each of the partners. Benefits to host plants can be positive (mutualism), neutral (commensalism and neutralism) or negative (parasitism, competition and amensalism). Variations in the outside environment puts the plant metabolism out of homeostasis which creates necessity for the plant to harbour some advanced genetic and metabolic mechanisms within its cellular system. Herein the importance of microbes, especially the endophytes increases immensely. They aid in plant health by deterring herbivory and pathogenesis while also facilitating plant growth through nutrients uptake (modification of root morphology, alteration of nitrogen accumulation and metabolism), water use efficiency (osmotic adjustment, stomatal regulation), and curtailing of environmental stresses. The endophytes, in return, obtain access to the host plant's nutrients and dissemination to the next generation.

## Abiotic Stress Tolerance in Plants via Endophytes

Symbiotically conferred abiotic stress tolerance involves at least two mechanisms:

- Activation of host stress response systems soon after exposure to stress, allowing the plants to avoid or mitigate the impacts of the stress.
- Biosynthesis of antistress biochemicals by endophytes.

Studies conducted on Arthrobacter sp. and Bacillus sp. isolated from pepper plant showed significant reduction in upregulation and even downregulation of some stress-inducible genes when compared with gene expression in uninoculated plants. Phomaglomerata-and Penicilliumsp. Significantly increased plant biomass, related growth parameters, assimilation of essential nutrients such as potassium, calcium, magnesium and reduced the sodium toxicity in cucumber plants under sodium chloride and polyethylene glycol–induced salinity and drought stress when compared with control plants.

## Enhancement of Water-use Efficiency during Drought

Endophyte associated plants (panic grass, rice, tomato and dune grass)have been reported to use significantly less water, increased biomassthan in non-symbiotic plants. Thedrought tolerance phenomenon may be explainedby enhanced accumulation of solutes in tissuesof endophyte-infected plants as comparedto non-infected plants, or by reducedleaf conductance and a slowdown of thetranspiration stream, or due to thicker cuticleformation. Characterization of Trichoderma/Theobromacacao revealed changes in gene expression patterns which imply the possibility that Trichoderma spp. could induce tolerance to abiotic stresses, possibly including drought, in cacao.

## Responses to Osmotic Stress

A wide range of responses are exhibited against osmotic stress by plants at the molecular, cellular and whole-plant level such as, inhibition of shoot growth and enhancement of root growth, adjustment in ion transport (uptake, extrusion and compartmentalization of ions) and metabolic changes (carbon metabolism, the synthesis of compatible solutes). While some of these responses are trigged by the primary osmotic stress signals, others may result from secondary stresses/signals caused by the primary signals. These secondary signals are phytohormones (e.g. ABA, ethylene), ROS and intracellular second messengers (e.g. phospholipids). Some of these secondary signals may not be confined to the primary stress sites such as the root. Root-derived ABA can ascend with transpiration flow to regulate stomatal aperture in leaves under drought.

## Nutrient Stress (Nutrient Uptake and Recycling)

Endophytes enable supply of macronutrients and micronutrients to their host. Plant root exudates are metabolized by nitrogen fixing bacteria and in turn provide nitrogen to the plant for amino acid synthesis. Endophytes promote plant growth by phosphate solubilization, indole acetic acid (IAA) cytokinins, gibberellin (GAs),siderophore production and the supply of essential vitamins. Choi et al have reported Pseudomonas spp.mediated phosphate solubilization in rice and wheat by producing giberrelic acid. P. indica and Azotobacter chroococcum improved uptake of mineral nutrients (especially Zn) in wheat.Studies have confirmed the role of endophytes in biodegradation of the litter of its host plants. They initially colonize the plants and facilitate the saprophytic microbes to act on through antagonism thereby increasing litter decomposition. Another study demonstrated that all endophytes have the ability to decompose organic components, including lignin, cellulose, and hemicellulose. This way they aid in nutrient cycling.

## Salinity and Alkalinity Tolerance

Tolerance of plants to salt stress is associated with the alleviation of antioxidant enzymes as suggested by earlier studies. The ROS scavengers include glutathione, ascorbate and tocopherol and the enzymes superoxide dismutases (SOD), catalases (CAT), ascorbate or thiol-dependent peroxidases (APX), glutathione reductases (GR), dehydroascorbate reductases (DHAR) and mono-dehydroascorbate reductases (MDHAR). They are involved in the removal of ROS either directly (SOD, CAT, APX) or indirectly via regeneration ascorbate and glutathione in the cell.P. indica induces salt tolerance in barleyby increasing the levels of antioxidants.On constant exposure to 500 mM NaCl solution (sea water levels to mimic exposure of plants in their native beach habitat), nonsymbiotic plants Leymus mollis (dunegrass) became severely wilted and desiccated within 7 days and were dead after 14 days. In contrast, symbiotic plants infected with Fusarium culmorum did not show wilting symptoms until they were exposed to 500 mM NaCl solution for 14 days.

## Temperature Stress Tolerance

While most of the available literature has pointed towards bitrophic interactions between microbes and plants influencing biotic and abiotic stress, there is also tantalizing evidence that these interactions can be influenced at the tritrophic level. As a notable example, the prevalence of viruses in fungi (mycoviruses) and their influence on mutualistic relationships with plants has attracted recent attention. The ability of the grass Dichanthelium lanuginosumto survive soil temperatures ranging between 38°C and 65°C in Yellowstone National Park was directly linked to an association with the fungus Curvularia protuberata and its mycovirus, Curvularia thermal tolerance virus (CThTV). The endophytic fungi increased heat tolerance in wheat in terms of height, weight of grain, as well as germination of second-generation seeds.

### Heavy Metal Stress Tolerance

The root-associated dark septate endophyte, Exophiala pisciphila of Zea mays analysed under increased soil Cd stress (0, 10, 50, 100 mg kg–1) showed enhanced antioxidant enzyme activity. Three key genes involved in uptake, detoxification and transport: namely ZIP was downregulated; PCS upregulated and MTP was upregulated upon inoculation with DSE and exposed to subsequent high Cd concentrations. Alteration in the levels of ACC by Pseudomonas and Gigaspora can alter the tolerance of heavy metals directly through the manipulation of plant ethylene levels.

## FUNGAL ENDOPHYTES FOR IMPROVEMENT OF DROUGHT STRESS TOLERANCE

The interaction of microbes with the plant can be traced back to the origin of plants. The early evolution of plants occurred in a diverse microbial world. Archaea, bacteria, fungi, and viruses had been evolving for billions of years. The most well-known plant–microbe interaction is the mutualism between mycorrhizal fungi and plant where both partners generally benefit from each other. However, under natural settings, plants form relationships with endophytic fungi and viruses which can be beneficial or harmful for the partners depending on host types and natural and environmental situations. Plants in natural systems and crop lands are simultaneously exposed to both biotic and abiotic stresses. Though stress research is mostly focusing on plant response to a particular environmental stress, research focusing to both biotic and abiotic stresses together has also been conducted. Unravelling the complex mechanisms of plant–microbe relations and their effects in abiotic stress tolerance in plants could potentially advocate novel tactics to boost the productivity of crops.

The term 'endophyte' refers to the fungi that live inside the plant intercellular and intracellular spaces for at least part of life cycle, causing no concurrent visible symptoms at any specific moment. This definition of endophyte is strictly operational and contextual since it takes into account the result of a specific fungus–host interaction only in a given time under the particular environmental settings, because symptomless endophytes can behave differently (e.g. as pathogens) under altered environmental conditions. The existence of fungal endophytes from fossil records suggests that endophyte–host associations may have evolved from the time of development of first higher plants on earth. It is thought that the majority, if not all plants, have one or more types of these endophytes and numerous endophytic species; in some cases, above a hundred can be found in a certain plant species. Fungal endophytes have been documented from healthy aerial tissues of conifers and grasses. Further, fungal endophytes have also been reported from marine algae, lichens, mosses and ferns, palms and pteridophytes. Fungal endophytes can be grouped into three basic ecological groups:

- Mycorrhizal fungi.
- Balancious or 'grass endophytes'.
- Non-balancious endophytes.

However, Brundrett separated mycorrhizal from endophytic interactions in that mycorrhizas have coordinated plant–fungus development and nutrient transfer at specialized interfaces. Later, Rodriguez et al. classified the endophytes under two major groups, viz. clavicipitaceous and non-clavicipitaceous on the basis of phylogeny and life history traits. Clavicipitaceous fungal endophytes are limited to certain grasses, while non-clavicipitaceous ones have a broad host range including both nonvascular and vascular plant species. In addition, recent reviews propose that members of the non-clavicipitaceous group can be segregated into three subgroups on the basis of host range, type of tissue infected, pattern spread, in planta infection and the establishment, diversity and benefits given to hosts. A diverse kind of relationships exists between the fungal endophytes and plant ranging from mutualistic, symbiotic and commensal to pathogens. However, the state of the interaction between endophyte and host may be transitory, and many factors could make changes in their mode of interaction. In symbiotic associations, balansiaceous endophytes with their hosts are commonly considered as being mutualistic even though some of them provide nothing to their hosts and can occasionally be antagonistic. Although most of the endophytes are regarded as being mutualistic with their hosts, some fungal endophytes may become pathogenic to plants, depending on the developmental stage of the partners, environmental conditions and plant defence reactions. Endophytic fungi have been known to play a vital role in plant growth, especially grasses; however, few reports have elucidated their symbiosis with crops. Recently, the ecological roles of some endophytes have been explained. In addition to providing nutritional benefits, fungal endophytes also confer significant physiological and ecological benefits, including protection from environmental stress as well as from an attack of pathogens and pests.

## Mechanisms of Endophyte-mediated Plant Drought Tolerance

Fungal endophytes have been shown to provide fitness benefit to plant when exposed to water-limiting conditions. Perhaps the most widely documented example of endophyte-mediated drought stress tolerance in plants is the enhanced drought tolerance of tall fescue and perennial ryegrass due to infection of the endophyte Neotyphodium coenophialum. Kane studied with the leaf-inhabiting endophyte Neotyphodium lolii to assess its potential benefits or harm in drought stress tolerance of native perennial ryegrass collections formerly obtained from the Mediterranean regions. Non-grass fungal endophytes have also been described to help plants alleviate drought stress. endophyte colonization can help improve abiotic stress tolerance such as drought in that host. It must be noted that endophytic symbiosis in plants does not always benefit the plant under drought or other abiotic stress conditions, and their interactions could cost for plants in terms of their ability to stand in stresses. Cheplick reviewed the role of fungal endophytes on potential drought tolerance and cited some studies where endophytes imparted no improvement in the host's ability to tolerate drought stress. For instance, Zaurov et al. inoculated fescue plants with Neotyphodium isolates collected from dissimilar hosts. They observed that some genotypic combinations affected negatively on plant mass, some had no effect and others increased plant biomass. Similarly, few combinations improved tolerance to soil aluminium; others have neutral or decreased tolerance compared to endophyte-free clones. This study revealed that genotype-specific interactions may increase or decrease or have no effect on plant adaptation and fitness. Thus, endophyte-mediated response to water stress is a complex phenomenon involving various metabolites and metabolic pathways. While the ability of fungal endophytes to provide drought tolerance in host plants has been described in many studies, the underlying mechanisms are incompletely characterized. In an effort to illuminate the underlying mechanism by which endophyte causes increased drought tolerance, researchers have reported few observations. Research so far studying the effect of endophyte on plant responses to drought stress have described certain physiological, biological and biochemical modifications such as:

- Increased growth and development.
- Enhanced osmotic balance.
- Increased gaseous exchange and water-use efficiency.
- Improved defence against oxidative damage.

When water-limiting conditions may improve, alleviate and recompense the harmful effects of water stress in endophyte-colonized (EC) plants. The present chapter aimed at outlining the recent advances in the study of improvement of drought tolerance by endophyte colonization in plant subjected to water stress.

## Endophyte-mediated Plant Growth Enhancement

Fungal endophytes have been shown to enhance growth and biomass of plants under water-limiting conditions. For example, inoculation of Fusarium culmorum and Curvularia protuberata resulted in higher biomass of drought-affected rice plants than non-inoculated plants. Endophytes Chaetomium globosum and Penicillium resedanum isolated from Capsicum annuum plants promoted shoot length and biomass of the host plants subjected to drought stress. Drought-challenged tomato plants showed higher root and shoot biomass when inoculated with class 2 fungal endophytes, including Alternaria sp. and Trichoderma harzianum. Inoculating a Trichoderma hamatum isolate caused increased higher root fresh weight, dry weight and water content, regardless of water availability in Theobroma cacao (cacao). The endophyte Piriformospora indica colonization in Chinese cabbage promoted root and shoot growth and lateral root development. Production of auxins by fungal endophytes is attributed to the increased growth of plants under stress. Also, stress-induced endogenous abscisic acid and the genes involved, such as zeaxanthin epoxidase, 9-cis-epoxycarotenoid dioxygenase 3 and ABA aldehyde oxidase 3, have been found to be significantly decreased in endophyte-colonized plants under stress, the effect of which could be comparable to that of the exogenous GA3 in terms of promoting plant growth and yield under stressed conditions by manipulating hosts' physiological processes. However, in some cases, it was recorded that endophytes do not show positive effects on host growth during drought stress, but they help with rapid recovery of host plant after water became available again.

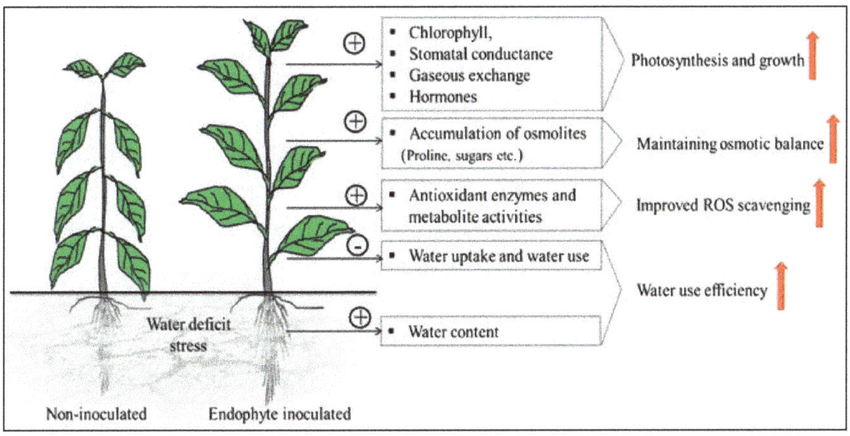


Endophyte colonization can help plants better withstand in water deficit stress by exerting their effects, directly or indirectly, on plant functions at both above and belowground. The plant on the left side which represents a drought-stressed non-colonized plant shows reduced growth and biomass due to lower photosynthetic rate, higher amount of oxidative damage, reduced uptake water and unbalanced osmoregulation. On the right side, a fungal endophyte-colonized plant under water deprivation stress is shown. Endophyte colonization shows increased growth and biomass due to enhanced photosynthetic rate, increased water-use efficiency and better osmotic balance. There is higher accumulation of osmolytes and lower degree of oxidative damage in the EC plants.

## Endophyte-mediated Improved Photosynthesis

Moisture stress causes decreased levels of photosynthesis in plants through decreased synthesis of ATP and other enzymes such as rubisco and sucrose–phosphate synthase as water availability decreases. Plant tolerance to water stress involves the management of extra radiation caused by reduced photosynthesis and $CO_2$ availability and a greater susceptibility to photo-damage. The endophyte-colonization results in higher chlorophyll content and leaf area in plants challenged by stress than non-colonized plant. Higher concentration of chlorophyll is associated with higher photosynthetic rate. The increased rate of photosynthesis was recorded from the droughtstressed Capsicum annuum plants colonized by endophytes Chaetomium globosum and Penicillium resedanum. About twofold increase in chlorophyll content and photosynthetic efficiency in P. indica-colonized Arabidopsis plants was measured when seedlings were challenged with waterlimiting conditions. P. indica reduced the drought-induced decline in the photosynthetic rate and the denaturation of chlorophyll and thylakoid proteins. Although, the Fv/Fm values decreased in the non-EC plants under drought, no significant difference was observed for the P. indicacolonized plants indicating that EC plants suffer less from water stress than uninoculated controls. In the same study, the total chlorophyll level was reported to be reduced by more than 50% in non-EC plant, but colonized plants showed only a slight decrease in total chlorophyll content. Additionally, a decrease in the protein levels of representative constituents of the thylakoid membrane and of enzymes situated in the plastid stroma in stressed plants was retarded when colonized with P. indica. Recently, Azad and Kaminskyj characterized a fungal endophyte that enhanced drought tolerance of the host and increased photosynthesis in the leaf. The mechanism of increased photosynthesis in EC plant under water stress is not fully understood. In one study, it was found that while the photosynthesis rate and stomatal conductance increased in droughtaffected EC plants, initial rubisco activity and carboxylation efficiency did not differ from non-EC plants. It was suggested that endophyte colonization might result in reduced biochemical damage to the photosynthetic machinery plants subjected to water stress.

## Plant – Water Relation and Osmotic Adjustment as Mediated by Endophyte

In the broad sense, decreasing water loss and maintaining water uptake are the key processes that plants employ to adapt to water-limiting environments. Maintaining water uptake is assisted within plant cells by osmotic adjustment (OA), a biochemical mechanism that helps plants to adapt to drought conditions. OA results in a net accrual of compatible solutes, also known as osmolytes in the cell so as to maintain the favourable gradient for water flow from soil into roots. This accumulation of various ions, amino acids and sugars leads to a more negative osmotic potential, which is important for maintaining cell hydration and turgor, cellular development and growth, stomatal opening, photosynthesis and water uptake during drought. Endophyte-colonized plants consume significantly less water than non-colonized plants. For example, significantly less water use has been reported in endophyte-inoculated panic grass, rice, tomato and dune grass, indicative of their more efficient water usage. Reduced water consumption and improved water-use efficiency may offer a distinctive mechanism for endophyte-mediated drought resistance in plants. Again, EC plants can maintain significantly greater water content than the noninoculated under water stress, implying the ability of endophytes to delay desiccation and damage in stress. The endophyte association could help plant access larger volumes of water from sources not reachable to the non-infected plants which suffered from stress. Endophyte association resulted in a decreased level of electrolytic leakage inside the plant tissues upon exposure to water deficit stress. Altered water potential and improved osmotic balance in drought-affected tall fescue infected with N. coenophialum endophyte have also been noted in some studies. Increased root water content was reported from T. hamatum-inoculated T. cacao plant subjected to water deficit stress compared to non-inoculated plants. A number of fungal endophytes have been reported to produce active biochemicals and metabolites that help the host plant withstand water deficit stress. Under drought conditions, significantly upregulation of free glucose, fructose, trehalose, sugar alcohols, proline and glutamic acid was detected in shoots and roots in tall fescue colonized by Neotyphodium coenophialum. Variable levels of proline accumulation were observed in EC plants subjected to water stress. While significantly more proline was accumulated in one genotype of tall fescue plant, no differences were observed in another genotype challenged with mimic drought in hydroponic culture when inoculated with Neotyphodium grass endophyte. Increased level of proline, soluble sugar and catalase (CAT) was observed in wheat colonized by endophyte Chaetomium globosum under water stress. Concentrations of aspartic acid and glutamic acid and of alanine and γ-aminobutyric acid were measured in drought-affected Theobroma cacao seedlings colonized by an isolate of Trichoderma hamatum. The changes in metabolites could be attributed to the strategies of EC plants towards drought tolerance or avoidance. Downregulation in osmolytes has previously been described as a strategy of drought avoidance, whereas the increase of osmoprotectants has been related to drought tolerance.

## Endophyte-mediated ROS Scavenging

Reactive oxygen species (ROS) act as signalling molecules in plants. ROS is involved in many plant processes, including growth, stress response, cell cycle and programmed cell death by influencing the expression of related genes. Abiotic stresses cause excess synthesis of these highly reactive molecules, these ROS causing oxidative stress and damaging proteins, lipids and DNA. Manufacturing additional ROS, i.e. hydrogen peroxide ($H_2O_2$), hydroxyl radical ($OH^{\cdot}$), singlet oxygen and superoxides ($^1O_2$), is one of the main mechanisms for plant cell damage or death in drought. Plants react against excess ROS through an intricate network of direct ROS-quenching activity or indirect hormone-mediated signalling activity. Various enzymatic and non-enzymatic antioxidant molecules are involved in scavenging ROS. Malfunctioning of these antioxidants' defence system results in oxidative damage in cells. Endophyte colonization simulates a more powerful ROS-scavenging system in host plants under stress and reduces damage of biomolecules at the cellular level. For instance, a lower level ROS production has been documented in endophyte-colonized tomato plants than in control plants following water stress. When plants were inoculated with P. indica and exposed to drought stress, up-regulation of peroxidase (POX), catalase (CAT) and superoxide dismutase (SOD) activities in the leaves was observed.

The level of another biomarker of oxidative stress, namely, malondialdehyde (MDA), was recorded to be lower in P. indica-colonized cabbage plants than in control plants. MDA is primarily produced through the ROS-induced degradation of polyunsaturated lipids. It is suggested that P. indica could prevent or reduce the damage of these lipids by inhibiting excess ROS production under stress conditions. Endophytes that promote drought tolerance have also been found to have high levels of loline alkaloids. Further experiments could test if these molecules are associated with the prevention of damage of macromolecules or reduction of ROS effects. Endophyteinduced production of antioxidant enzyme in plants under stress is predominantly observed in leaves. All these studies demonstrate that endophyte inoculation results in a strong defence response in plant in water stress, in which alleviation of oxidative stress might be a vital part. The study of nonvolatile compounds has been the major focus in most plant antioxidant research. However, plant leaves emitting volatile organic compounds could also play as a further defence system against stresses. The effect of volatile compounds such as isoprenoids has been described, where these compounds act as protective agent against oxidative stress in plant through direct ROS scavenging and indirect alteration of ROS signalling in arbuscular mycorrhizal plants. Endophyte-colonized plants could emit similar volatile organic compounds to cope with abiotic stress.

## References


- Role-of-plant-transcription-factors-in-abiotic-stress-tolerance, abiotic-stress-response-in plants-physiological-biochemical-and-genetic-perspectives: intechopen.com, Retrieved 16 July, 2020
- Induction-of-abiotic-stress-tolerance-in-plants-by-endophytic-microbes-322658634: researchgate.net, Retrieved 25 August, 2020
- The-role-of-plant-growth-promoting-bacteria-in-the-growth-of-cereals-under-abiotic-stresses, root-biology-growth-physiology-and-functions: intechopen.com, Retrieved 25 April, 2020
- Plant-Fungi-Association-Role-of-Fungal-Endophytes-in-Improving-Plant-Tolerance-to-Drought-Stress-318440889: researchgate.net, Retrieved 05 July, 2020
- Soil-bacteria-support-and-protect-plants-against-abiotic-stresses, abiotic-stress-in-plants-mechanisms-and-adaptations: intechopen.comm, Retrieved 17 March, 2020

# Breeding for Stress Tolerance

CHAPTER 5

Breeding in plants is the process of development of new varieties of plants that provides stress tolerance against drought, heat, water, cold temperature, etc. This chapter closely examines the concepts associated with breeding for stress tolerance to provide an extensive understanding of the subject.

## ENGINEERING COLD STRESS TOLERANCE IN CROP PLANTS

Plants respond with changes in their pattern of gene expression and protein products when exposed to low temperatures. Thus ability to adapt has an impact on the distribution and survival of the plant, and on crop yields. Many species of tropical or subtropical origin are injured or killed by non-freezing low temperatures, and exhibit various symptoms of chilling injury such as chlorosis, necrosis, or growth retardation. In contrast, chilling tolerant species are able to grow at such cold temperatures. Conventional breeding methods have met with limited success in improving the cold tolerance of important crop plants involving inter-specific or inter-generic hybridization. Recent studies involving full genome profiling/sequencing, mutational and transgenic plant analyses, have provided a deep insight of the complex transcriptional mechanism that operates under cold stress. The alterations in expression of genes in response to cold temperatures are followed by increases in the levels of hundreds of metabolites, some of which are known to have protective effects against the damaging effects of cold stress. Various low temperature inducible genes have been isolated from plants. Most appear to be involved in tolerance to cold stress and the expression of some of them is regulated by C-repeat binding factor/dehydration-responsive element binding (CBF/DREB1) transcription factors. Numerous physiological and molecular changes occur during cold acclimation which reveals that the cold resistance is more complex than perceived and involves more than one pathway.

Plants respond with changes in their pattern of gene expression and protein products when exposed to low temperatures. Thus ability to adapt has an impact on the distribution and survival of the plant, and on crop yields. Many species of tropical or subtropical origin are injured or killed by non-freezing low temperatures, and exhibit various symptoms of chilling injury such as chlorosis, necrosis, or growth retardation. In contrast, chilling tolerant species are able to grow at such cold temperatures. Conventional breeding methods have met with limited success in improving the cold tolerance of important crop plants involving inter-specific or inter-generic hybridization. Recent studies involving full genome profiling/sequencing, mutational and transgenic plant analyses, have provided a deep insight of the complex transcriptional mechanism that operates under cold stress. The alterations in expression of genes in response to cold temperatures are followed by increases in the levels of hundreds of metabolites, some of which are known to have protective effects against the damaging effects of cold stress. Various low temperature inducible genes have been isolated from plants. Most appear to be involved in tolerance to cold stress and the expression of some of them is regulated by C-repeat binding factor/dehydration-responsive element binding (CBF/DREB1) transcription factors. Numerous physiological and molecular changes occur during cold acclimation which reveals that the cold resistance is more complex than perceived and involves more than one pathway. The findings summarized in this review have shown potential practical applications for breeding cold tolerance in crop and horticultural plants suitable to temperate geographical locations.

Abiotic stresses adversely affect growth, productivity and trigger a series of morphological, physiological, biochemical and molecular changes in plants. Cold stress is a major environmental factor that limits the agricultural productivity of plants in hilly areas. Plants respond and adapt to this stress to survive under stress conditions at the molecular and cellular levels as well as at the physiological and biochemical levels. However, expression of a variety of genes is induced by different stresses in diverse plants.

Low temperature often affects plant growth and crop productivity, which causes significant crop losses. Plants differ in their tolerance to chilling (0-15 °C) and freezing (< 0 °C) temperatures. In general, plants from temperate climatic regions are considered to be chilling tolerant with variable degree, and can increase their freezing tolerance by being exposed to chilling, non-freezing temperatures, a process known as cold acclimation, which is associated with biochemical and physiological changes and ultimately showed marked changes in gene expression, biomembrane lipid composition, and small molecule accumulation. Besides, plants of tropical and subtropical origins. are sensitive to chilling stress and lack the mechanism of cold acclimation. Low temperature resistance in plants is a very complex trait, involving many different metabolic pathways and cell compartments. Conventional breeding methods have met with limited success in improving the cold tolerance of important crop plants involving inter-specific or inter-generic hybridization. Besides, in vitro induced variations have also been applied to improve the abiotic stress tolerance of various crop plants but without much success. The conventional breeding approaches are limited by the complexity of stress tolerance traits, low genetic variance of yield components under stress condition and lack of efficient selection criteria. It is important, therefore, to look for alternative strategies to develop cold stress tolerant crops.

Biotechnology offers new strategies that can be used to develop transgenic crop plants with improved tolerance to cold stress. Rapid advance in recombinant DNA technology and development of precise and efficient gene transfer protocols have resulted in efficient transformation and generation of transgenic lines in a number of crop species. A number of genes have been isolated and characterized that are responsive to freezing stress. Many studies have suggested that cold regulated gene expression is critical in plants for both chilling tolerance and cold acclimation. Advent of molecular tools has made it possible to select directly at the gene label without waiting for the phenotype to show up. Transgenic approach is being pursued actively throughout the world to improve traits including tolerance to biotic and abiotic stresses in a number of crops. As compared to other stresses, plant responses to cold stress are complex, so the prospects of improving cold tolerance in crops seem not to be very bright. Despite this, efforts have been made during the last two decades to generate transgenic lines of different crops, which have shown improved tolerance to cold stress. Therefore, it is important to use most appropriate tools that help in reaching the goals. The genotype designed should be better than the available ones and must reach the farmers.

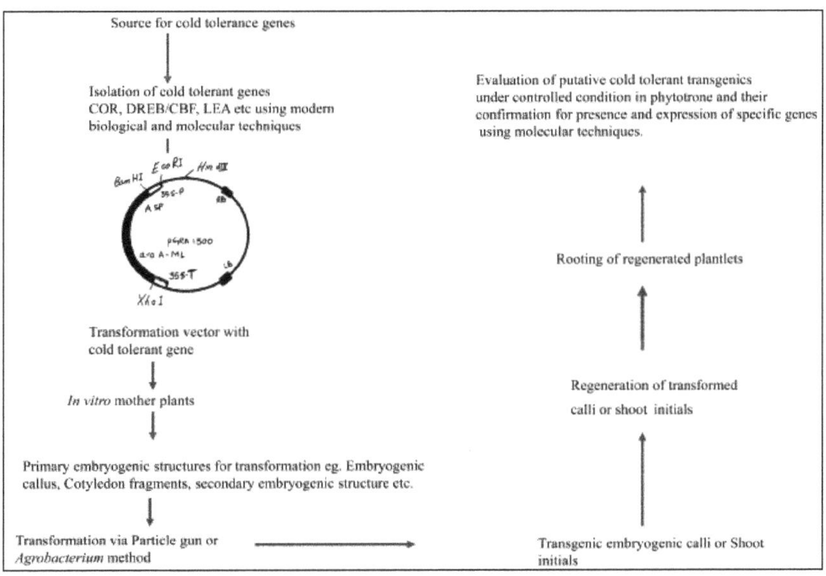


Scheme for genetic engineering of crop plants against cold stress.

## Morpho-physiological basis of Cold Tolerance

A large number of studies have evaluated different plant species tolerant to different stresses such as drought, salinity and cold. However, less detail is given with regard to the methods used to evaluate the stress response; these studies may bring about some misleading conclusions from an agronomic or physiology perspective. This is particularly important, in order to closely mimic the life span of most crops under cycles of stress, rather than short exposure to very severe stresses; although we agree that short exposures to stress are certainly adequate if the purpose is to assess gene expression only.

The temperate and cool regions are those where altitudes ranged from 1600-2500 m amsl (above mean sea level) and temperature during crop growth period ranged from $5-20\,^{\circ}C$. In temperate regions, low temperature is the primary abiotic stress which limits the crop productivity. The low temperature at seedling and reproductive stages is the major problem, results in slow establishment and low seed set which leads to poor yield of the crop. The low temperature limits the crop productivity when temperatures remain above freezing that is >0 °C, it is called as chilling stress. Chilling sensitive cultivars are typically tropical genotypes. There is wide range of cold stress in temperate areas differing in both timing and intensity of low temperature. Yield losses are more severe when cold stress occurs during reproductive stage/anthesis in rice which lead to high spikelet sterility. Ability of crop genotypes/lines to survive/perform better under low temperature than other genotypes is called as cold tolerance. Ordinarily, it is the consequence of cold hardening that is an earlier exposure to a low temperature for a specific period as a result of which chilling tolerance of the concerned plants increases. Cold tolerance involves increased chlorophyll accumulation, reduced sensitivity of photosynthesis, improved germination, pollen fertility and seed set.

## Increased Chlorophyll Accumulation

Low temperature inhibits chlorophyll accumulations in actively growing leaves. In rice, cold tolerant lines, for example, japonica accumulates more chlorophyll under cold stress than do cold sensitive line, for example of indica rice. Rasolofo evaluated 181 accessions to identify donor and outstanding cold tolerant lines using leaf discolouration score (1-3) and found 19 remained green (dark) after 10 days in the 12 °C cold water tank. Sanghera et al. found 18 cold tolerance IRCTN rice genotypes based on dark green colour and high spikelet fertility (>90%) under temperate conditions.

## Reduced Sensitivity of Photosynthesis

Chloroplast and photosynthesis is major site of cold injury. Tolerance in these aspects is expressed in native vegetation adapted to growing under cool conditions. The reduced sensitivity of photosynthesis to cold has been observed in maize inbreds adapted to low temperature which is partly related to specific enzymes of the process.

## Improved Germination

Genetic variation in cold tolerance at germination and seedling stage has been documented. Saini and Tandon found that L62G, Heng Jodo, Jodo, Heugdo, IRAT 102, Khonorullo, K 78-28, Daegaldo, Mujudo and L62-2A genotypes were cold tolerant having more than 85% germination and good seedling vigour (score 3) at an average of 11 °C field temperature

## Improved Pollen Fertility and Seed Set

Cold tolerance at reproductive stage is expressed as improved seed set and pollen fertility. It is largely a function of floral structure and function under stress. Lia et al. reported plant cold tolerance in rice is associated with anther size, number of pollen grain, diameter of fertile pollen grains at booting stage. However, Sanghera et al. reported that cold tolerance is associated with high spikelet fertility (>90%) and well panicle exsertion under temperate conditions.

Cold snaps cause a reaction in the plant that prevents sugar getting to the pollen. Without sugar there is no starch build-up which provides energy for pollen germination. And without pollen, pollination cannot occur, thereby, no grain is produced. CSIRO has found that all the ingredients for starch are present but they are not getting into the pollen grain where they are needed. A cell layer surrounding the pollen, called the 'tapetum', is responsible for feeding the pollen with sugar. The tapetum is only active for 1-2 days – so if a cold snap occurs at this time, then there is no further chance for pollen growth. But the sugar cannot freely move into the tapetum and pass through it to the pollen. Instead the sugar has to be broken down then transported in bits to the pollen. 'Invertase' is the catalyst that helps in breakdown of the sugar molecule to transport it into the tapetum before it is transported to the pollen. Quantities of invertase are decreased in conventional rice when it is exposed to cold temperatures, but they remain at normal levels in a cold tolerant variety when it experiences cold. By comparing a cold tolerant strain of rice with conventional rice, CSIRO has found that the gene responsible for invertase looks exactly the same in the cold tolerant variety as it does in conventional rice. So the invertase gene itself does not make the rice plant cold tolerant – but instead a mechanism that regulates the invertase gene is different. Early research indicates that the invertase gene is regulated by the hormone abscisic acid (ABA). Oliver et al. has experimented with injecting plants with ABA – the resulting rice plants are sterile, just like that if they experienced a cold snap. Also, ABA levels increase when conventional rice is exposed to cold, but they remain the same in the cold tolerant variety. Recent studies have indicated that the difference between cold-sensitive and tolerant rice is due to a different ability to control ABA levels. It has also been shown that this mechanism may require interactions with other plant hormones like auxins. Further, Zhao et al. also reported that low temperature turns off the genes responsible for sugar transport into the pollen grains and therefore starch cannot be produced in the pollen in cold conditions. Cold did not cause repression of sugar delivery in cold tolerant Chinese rice and fertile pollen was still produced following cold treatment. The sugar metabolism genes also continued to function normally during cold treatment of cold tolerant rice. Ample genetic variation for cold tolerance is available in well adapted breeding population. Germplasm collected from high altitude and low temperature areas, cold tolerant mutant, somaclonal variants and wild species can be exploited for breeding improved cold tolerant genotypes in hilly areas.

## Mechanisms for Understanding Tolerance to Cold Injury

Low temperature has a huge impact on the survival and geographical distribution of plants. It affects a range of cellular metabolisms in plant system depending on the intensity and duration of the stress. When exposed to low temperatures, plants respond with changes in their pattern of gene expression and protein products. Different studies have indicated that the membrane systems of the cell are the primary site of freezing injury in plants. In addition, it is well established that freeze-induced membrane damage results primarily from the severe dehydration associated with freezing. Many species of tropical or subtropical origin are injured or killed by nonfreezing low temperatures, and exhibit various symptoms of chilling injury such as chlorosis, necrosis, or growth retardation. In contrast, chilling-tolerant species are able to grow at such cold temperatures.

However, multiple forms of membrane damage can occur as a consequence of freeze induced cellular dehydration including expansion-induced-lysis, lamellar-to-hexagonal-II phase transitions, and fracture jump lesions. Thus, a key function of cold acclimation should be to stabilize membranes against freezing injury. Indeed, cold acclimation prevents expansion-induced-lyses and the formation of hexagonal II phase lipids in rye and other plants. This ability to adapt has an impact on the distribution and survival of the plant, and on crop yields. Multiple mechanisms appear to be involved in this stabilization. The best documented are changes in lipid composition. Secondly, temperature induced change in membrane fluidity is another consequences in plants during temperature stresses and might represent a potential site of perception and injury. Adaptation of living cells to chilling temperatures is a function of alteration in the membrane lipid composition by increased fatty acid unsaturation. Genetically engineered tobacco plants over-expressing chloroplast glycerol-3-phosphate acyltransferase (GPAT) gene (involved in phosphatidyl glycerol fatty acid desaturation) from squash (Cucurbita maxima) and A. thaliana showed an increase in the number of unsaturated fatty acids and a corresponding decrease in the chilling sensitivity. At low temperature, greater membrane lipid unsaturation appears to be crucial for optimum membrane function. An Arabidopsis fatty acid biosynthesis (FAB1) mutant with more saturated membranes showed decreased quantum efficiency of photosystem II (PSII), chlorophyll content and the amount of chloroplast glycerolipids after prolonged exposure to low temperature. A triple mutant fatty acid desaturation (fad3-2 fad7-2 fad8) devoid of trienoic fatty acids (18:3 or 16:3) produced a phenotype similar to FAB1, when plants were subjected to prolonged low temperature exposure. Similarly, fad5 and fad6 mutants with more saturated membranes became chlorotic and showed growth retardation during low temperature incubation. In addition to membrane unsaturation, it appears that lipid asymmetry in the membrane also contributes to membrane physical structure at low temperature. The accumulation of sucrose and other simple sugars that typically occurs with cold acclimation also seems likely to contribute to the stabilization of membranes as these molecules can protect membranes against freeze-induced damage in vitro.

Besides, protective chaperone like function of LEA proteins acting against cellular damage has been proposed, indicating the role of LEA proteins in anti-aggregation of enzymes under desiccation and freezing stresses. There is emerging evidence that certain novel hydrophilic and late embryogenesis abundant (LEA) polypeptides also participate in the stabilization of membranes against freeze-induced injury. These hydrophilic and late embryogenesis abundant polypeptides are predicted to contain regions capable of forming amphipathic α-helices which are shown to have strong effect on intrinsic curvature of monolayers and their propensity to form hexagonal II phase. They are said to defer their formation at lower temperatures. There is another evidence that freeze-induced production of reactive oxygen species contributes to membrane damage and that intercellular ice can form adhesions with cell walls and membranes and cause cell rupture. Further, there is evidence that protein denaturation occurs in plants at low temperature which could potentially result in cellular damage. In these cases, the enhancement of antioxidative mechanisms, increased levels of sugars in the apoplastic space, and the induction of genes encoding molecular chaperones, respectively, could have protective effects.

In cold stress-tolerant plants, many genes involved in the synthesis of osmoprotectants—organic compounds such as amino acids (e.g. proline), quaternary and other amines (e.g. glycinebetaine and polyamines) and a variety of sugars and sugar alcohols (e.g. mannitol, trehalose and galactinol) that accumulate during osmotic adjustment—have been used. Both cold-stress-induced transcripts and constitutively expressed transcripts need to be processed, exported to the cytoplasm and kept in conformations that are competent for translation. RNA can fold into extensive secondary structures that could interfere with its function, and cold temperatures exacerbate this interference. Many genes that respond to multiple stresses like dehydration and low temperature at the transcriptional level are also induced which protects the cell from dehydration and chilling. In order to restore the cellular function and make plants more tolerant to stress, transferring a single gene encoding a single specific stress protein may not be sufficient to reach the required tolerance levels. To overcome such constraints, enhancing tolerance towards multiple stresses by a gene encoding a stress inducible transcription factor that regulates a number of other genes is a promising approach. In bacteria, nucleic-acid-binding cold shock proteins (CSPs) accumulate at cold temperatures and function as transcription antiterminators or translational enhancers by destabilizing RNA secondary structure. Some CSP-domain-containing proteins in plants are upregulated by cold stress, and might function as RNA chaperones in the regulation of translation. A different cold-responsive nucleic-acid-binding protein, a zincfinger containing glycine-rich RNA-binding protein from Arabidopsis designated atRZ-1a, is also upregulated by cold stress, and genetic analysis supports its function in freezing tolerance. Compared to other organisms, plants have the largest number of DEAD-box RNA helicase genes. One of these helicases, which is encoded by the Arabidopsis low expression of osmotically responsive genes4 (LOS4) gene, is essential for plant tolerance of chilling and freezing stress. LOS4 is required for efficient export of RNA from the nucleus to the cytoplasm. The Arabidopsis nucleoporin AtNUP160 suppressor of auxin resistance1 (SAR1) also controls RNA export, and is crucial for chilling and freezing tolerance. Both LOS4 and AtNUP160 proteins are enriched at the nuclear rim. Defects in the nucleocytoplasmic transport of RNA seem to affect cold tolerance preferentially, because the LOS4 and AtNUP160 mutant plants do not have severe growth or developmental phenotypes, nor are they strongly altered in the tolerance of other abiotic stresses.

## Key Players Involved in Cold Responsive Pathways

Cold tolerance is the result of complex physiological mechanisms involving many cell and plant traits. Earlier studies have shown that the genetic control of cold tolerance is complex and can be regarded as polygenic and the mechanism of how these genes controlled cold tolerance is still not fully clear. Therefore, the foundation for a better molecular and genetic understanding of the cold responsive pathways will improve our knowledge and pave the way for the development of improved methodologies for cold tolerance screening. Key to the tolerance of plants to abiotic stresses is a complex network of transcription factors and other regulatory genes that control multiple defense enzymes, proteins and pathways. The discovery of gene expression change during cold acclimation was the starting of exploration of antifreezing molecular mechanisms. In this context, Zhao et al. reported that gene expression profiling using DNA chips indicated that large numbers of genes were differentially expressed under cold stress. They identified 242 unique genes that are expressed differentially between cold sensitive and cold tolerant rice. These genes are involved in processes such as senescence, cell death, male sterility and plant hormone response. Similarly, global transcript profiling analyses indicated that >10% of genes in the Arabidopsis genome were regulated during cold acclimation. Transcriptome analysis using microarray technology is a powerful technique, which has proven very useful for discovering many stress-inducible genes involved in stress response and tolerance. Genes involved in stress signal sensing and a cascade of stress-signaling in A. thaliana has been of recent research interest. Components of the same signal transduction pathway may also be shared by various stress factors such as drought, salt and cold. Although there are multiple pathways of signal-transduction systems operating at the cellular level for gene regulation.

In past, it has been reported that genes induced during stress conditions function not only in protecting cells from stress by producing important metabolic proteins, but also in regulating genes for signal transduction in the stress response. Several stress induced cor genes such as rd29A, cor15A, kin1 and cor6.6 are triggered in response to cold treatment, ABA and water deficit stress in the early stages of the osmotic stress response.

Several stress induced cor genes such as rd29A, cor15A, kin1 and cor6.6 are triggered in response to cold treatment, ABA and water deficit stress in the early stages of the osmotic stress response. Similarly, a cis-acting element, dehydration responsive element (DRE) identified in A. thaliana, is also involved in ABA-independent gene expression under drought, low temperature and high salt stress conditions in many dehydration responsive genes like rd29A that are responsible for dehydration and cold-induced gene expression. Thus, clearly, the overexpression of some drought-responsive transcription factors can lead to the expression of downstream genes and the enhancement of abiotic stress tolerance in plants. Thus, these gene products are classified into two groups. The first group of proteins that probably function in stress tolerance includes chaperones, LEA proteins, osmotin, antifreeze proteins, mRNA-binding proteins, some key enzymes for osmolyte biosynthesis (like proline, water channel proteins, sugar and proline transporters, detoxification enzymes), enzymes for fatty acid metabolism (proteinase inhibitors, ferritin) and lipid-transfer proteins. It has been reported that some of these stress-inducible genes specially encoding proteins (such as enzymes for osmolyte biosynthesis, LEA proteins and detoxification enzymes) have been overexpressed in transgenic plants and produce stress-tolerant phenotypes in the transgenic plants. These results indicate that the gene products of the stress-inducible genes really function in stress tolerance. The second group contains protein factors involved in regulation of signal transduction and gene expression that probably function during stress response. This group includes various transcription factors that regulate different stress-inducible genes collectively or separately, and may constitute gene networks. Seki et al. reported that some of these regulatory pathways are also involved in drought, cold, or high-salinity stress responses. Though, the clear cut functions of most of these genes are not fully understood. Functional analysis of these stress-inducible transcription factors will provide precise information on the complex regulatory gene networks that are involved in responses to drought, cold, and high-salinity stresses. Some of these stress-inducible regulatory genes that encode proteins (transcription factors) have been overexpressed in transgenic plants and generate stress-tolerant phenotypes in them.

## Cold Tolerance using Transgenic Approaches

When a plant is subjected to abiotic stress, a number of genes are turned on, resulting in increased levels of several metabolites and proteins, some of which may be responsible for conferring a certain degree of protection to these stresses. A key to progress towards breeding better crops under stress has been to understand the changes in cellular, biochemical and molecular machinery that occur in response to stress. A key to progress towards breeding better crops under stress has been to understand the changes in cellular, biochemical and molecular machinery that occur in response to stress which in turn provides new tools and strategies to improve the environmental stress tolerance of crops. Since freezing tolerance is a multigenic trait, transformation of a single functional gene appears to have a limited effect on crop freezing tolerance. Because many aspects of cold adaptation process are under transcriptional control, many transcription regulatory factors were chosen, hence, genetic engineering for introgression of such genes that are known to be involved in stress response and putative tolerance, might prove to be a faster track towards improving crop varieties for enhanced cold tolerance.

Low-temperature limitations have been overcome by the identification of cold-tolerant genes for applications in genetically transformed crops. In transgenic tobacco (Nicotiana tabacum), chilling tolerance at 1°C for 7 d was achieved by the over expression of a gene encoding chloroplast w3 fatty acid desaturase. Furthermore, tolerance at 1°C for 11 d was conferred using a gene encoding a non-specific cyanobacterial desaturase, and the resultant transgenic tobacco plants showed a reduction in saturated fatty acid content in membrane lipids. The over expression of glycerol-3-phosphate acyl transferase altered the unsaturation of fatty acids and conferred chilling tolerance in transgenic plants. Hence, modifications in lipid composition that stabilize cell membranes and prevent cellular leakage lead to cold tolerance.

Transgenic technology has opened up many exciting possibilities to improve cold stress in plants by introduction or removal of gene or genes that regulate a specific trait. It also offers uncommon opportunities for improvement in genetic potential of plants in the form of development of specific crop varieties that are more resistant to biotic and cold stresses with enhanced nutritional level.

During last two decades, advancement in plant biotechnology has led to the identification and isolation of a number of transcription factors related to cold stress tolerance. A good number of genes which have been identified in different studies raise the question of exactly which genes are most central to increasing cold/freezing tolerance. The genes selected for transformation should be involved in encoding enzymes that are required for the biosynthesis of various osmoprotectants. Other classes of genes that selected for transformation include those that encoded enzymes for modifying membrane lipids, LEA protein, and detoxification enzymes. In these studies, either a single gene for a protective protein or an enzyme was overexpressed under the control of the constitutive 35S cauliflower mosaic virus (CaMV) promoter in transgenic plants, although several genes have been shown to function in environmental stress tolerance and response. The genes encoding protein factors that regulate gene expression and signal transduction, that function in stress responses may be useful for improving the cold tolerance of plants by gene transfer as they can regulate many stress-inducible genes involved in cold stress tolerance.

The CBF genes represent one of the most significant discoveries in the field of low temperature adaptation and signal transduction. All important crops and few vegetables species have contained this gene. Various low temperature-inducible genes have been isolated from plants that appear to be involved in tolerance to cold stress, and the expression of some of them is regulated by C-repeat binding factor/dehydration-responsive element binding (CBF/ DREB1) transcription factors. Three CBF/DREB1 genes (CBF3/DREB1a, CBF1/DREB1b, and CBF2/DREB1c) belonging to the AP2/DREBP family of DNA-binding proteins have been identified in Arabidopsis. Transgenic Arabidopsis plants constitutively over-expressing a cold inducible transcription factor (CBF1; CRT/DRE binding protein) showed tolerance to freezing without any negative effect on the development and growth characteristics. Overexpression of Arabidopsis CBF1 (CRT/DRE binding protein) has been shown to activate cor homologous genes at non-acclimating temperatures. The CBF1 cDNA when introduced into tomato (Solanum lycopersicum) under the control of a CaMV35S promoter improved tolerance to chilling, drought and salt stress but exhibited dwarf phenotype and reduction in fruit set and seed number per fruit.

The expression of related cold shock proteins (CSPs) from bacteria, CspA from Escherichia coli and CspB from Bacillus subtilis, promotes stress adaptation in multiple plant species. Transgenic rice plants expressing CspA and CspB manifest improved stress tolerance for a number of abiotic stresses, including cold, heat, and water deficits. It has long been established that changes in gene expression occur upon exposure to cold acclimation. Number of COR genes isolated from Arabidopsis that encode polypeptides thought to have protective roles against dehydration. Expression profile experiments in Arabidopsis demonstrated that extensive changes in gene expression occur during cold acclimation and that a substantial number of the genes that are up-regulated by the cold response are involved in metabolism. Hajela et al. reported that a set of genes that encode a related family of cold-regulated (COR) proteins, which are massively induced during cold acclimation. These COR genes were used by different researchers to identify a family of Arabidopsis transcription factors known as either C-repeat binding factors (CBF) (CBF1, CBF2 and CBF3) or dehydration responsive element-binding factors (DREB) (DREB1B, DREB1C and DREB1A). In Arabidopsis (A. thaliana) and rice, the CBF/DREB1-dependent cold response pathway has been shown to play a predominant role in freezing tolerance through the process of cold acclimation. These CBFs/DREBs are upstream transcription factors that bind to promotor cis element CRT/DRE and activate the expression of these cold responsive genes. Overexpression of CBF1/DREB1b and CBF3/DREB1a enhances cold tolerance by inducing COR (cold regulated) genes. Furthermore, its overexpression causes many biochemical changes, such as the accumulation of sugar and proline. Thus, the CBF/DREB1 genes are thought to be activators that integrate several components of the cold acclimation response by which plants increase their tolerance to low temperatures after exposure to nonfreezing conditions. Several stress induced COR genes such as rd29A, COR15A, kin1 and COR6.6 are triggered in response to cold treatment, ABA and water deficit stress.

There have been numerous efforts in enhancing tolerance towards multiple stresses such as cold, drought and salt stress in crops other than the model plants like Arabidopsis, tobacco and alfalfa. An increased tolerance to freezing and drought in Arabidopsis was achieved by overexpressing CBF4, a close CBF/DREB1 homolog whose expression is rapidly induced during drought stress and by ABA treatment, but not by cold. Subsequently, the overexpression of DREB1A has been shown to improve the drought and low-temperature stress tolerance in tobacco, wheat and groundnut. The use of stress inducible rd29A promoter minimized the negative effects on plant growth in these crop species. Further, Gilmour et al. have shown that ectopic transgenic overexpression of CBF1/DREB1B, CBF2/ DREB1C or CBF3/DREB1A in Arabidopsis activates a suite of CBF/DREB target genes at warm temperatures and results in increased freezing and drought tolerance. CBF transcripts begin accumulating within 15 minutes of plants being exposed to low temperature strongly suggests that the low temperature "Thermometer" and "Signal Transducer" are present at warm non-inducing temperatures. Two cDNA clones that encode DRE-binding proteins, DREB1A/CBF3 and DREB2A, have been isolated by using yeast one-hybrid screening. cDNA clones encoding two DREB1A homologs (named DREB1B/CBF1 and DREB1C/CBF2) and one DREB2A homolog (DREB2B) were also isolated and Expression of the DREB1A gene and its two homologs was induced by low-temperature stress, whereas expression of the DREB2A gene and its homolog was induced by dehydration. These results indicate that two independent families of DREB proteins, DREB1 and DREB2, function as transacting factors in two separate signal transduction pathways under low-temperature and dehydration conditions, respectively.

However, CBFs are negatively regulated by an upstream transcription factor, MYB15 (an R2R3-MYB family protein) in Arabidopsis. MYB15 is expressed even in the absence of cold stress, and MYB15 can bind to MYB recognition elements (MYBRS) in the promoters of CBFs. MYB15 mutant plants show enhanced expression of CBFs during cold acclimation and enhanced freezing tolerance, whereas, transgenic Arabidopsis overexpressing MYB15 showed a decreased expression of CBFs and a reduction in freezing tolerance. Thus, MYB15 is an upstream transcription factor that negatively regulates the expression of CBFs.

Transcriptome analysis of ZAT12-overexpressing Arabidopsis revealed that the ZAT12 regulon consists of at least 24 cold standard set (COS) genes, of which nine are cold-induced and 15 are cold-repressed genes. Constitutive overexpression of Arabidopsis DREB1A improved drought and low-temperature stress tolerance in tobacco, and regulation of transgene expression via the stress-inducible RD29A promoter minimized the negative effects on plant growth. Similarly, the Arabidopsis DREB1A gene was placed under control of the RD29A promoter and transferred via biolistic transformation into bread wheat. However, constitutive overexpression of the CBF genes using the cauliflower mosaic virus 35S promoter can result in undesirable agronomic traits. In Arabidopsis, CBF overexpression can cause a "stunted" growth phenotype, a decrease in seed yield and a delay in flowering.

In the last decade, extensive research efforts have been undertaken to identify and characterize cold-responsive (COR) genes and a number of homologous components of the Arabidopsis CBF cold response pathway in many plants have been found. Many of these putative orthologs have been structured, analyzed and functionally tested. The expression patterns of the CBFs and CORs in response to low temperature are similar in a variety of plants species, involving rapid cold-induced expression of the CBFs followed by expression of CBF-targeted genes that increase freezing tolerance. Moreover, constitutive overexpression of the Arabidopsis CBF genes in other plants resulted in increased freezing tolerance that have been successfully used to engineer cold stress tolerance in several crop species.

Transgenic attempts with other structural genes have also been made with fair degree of success. Genetically engineered tobacco plants over-expressing chloroplast glycerol-3-phosphate acyltransferase (GPAT) gene (involved in phosphatidyl glycerol fatty acid desaturation) from squash (Cucurbita maxima) and A. thaliana showed an increase in the number of unsaturated fatty acids present in the plant cell wall, which enhance the cold tolerance to the plants during cold stress. Expression of a plant phosphatase (At PP2CA) in transgenic A. thaliana can accelerate the development of cold acclimation and increase freezing tolerance. It has also shown that transgenic plants expressing a constitutively active kinase NPK1, is more tolerant to chilling and other abiotic stresses. Pennycooke et al. reported that down-regulating α-Gal (α-Galactosidase) in transgenic petunia resulted in an increase in freezing tolerance suggesting that engineering raffinose metabolism by transformation with α-Gal provides an additional method for improving the freezing tolerance of plants. The overexpression of genes encoding LEA proteins can improve the stress tolerance of transgenic plants. Expression of the citrus gene encoding a LEA protein, CuCOR19 increased the cold tolerance of transgenic tobacco. Likewise, the freezing tolerance of Arabidopsis was increased by the ectopic expression of the wheat gene WCS19, the Arabidopsis gene COR15A, and the co-expression of the genes RAB18 and COR47 and XERO2 and ERD10. The freezing tolerance of strawberry leaves was enhanced by expression of the wheat dehydrin gene WCOR410. On the other hand, the expression of two cold-induced LEA proteins from spinach and three desiccation-induced LEA proteins from C. plantagineum in tobacco did not induce any significant changes in the freezing or drought tolerance of the respective transgenic plants. This may indicate either that not all LEA proteins make a significant contribution to plant stress tolerance, or that they need a particular background to function in, as suggested for transgenic strawberry plants. Kim et al. engineered tobacco with ring zinc finger protein (RDCPt) from hot pepper (Capsicum annuum) and reported that expression of this gene resulted in improved cold tolerance in transgenic plants as compared to wild type. Dai et al. reported that overexpression of OsMYB3R-2 in transgenic Arabidopsis increased tolerance to freezing when exposed to -8 °C for 10 h. They found that survival after 6 d at normal conditions was 26.8% for the wild-type and 84.5% for transgenic lines. Phenotypically, most transgenic seedlings were green and could regrow as compared with the wild type; whereas most wild-type seedlings became white and did not regrow after removed to normal conditions. The survival percentage under different low temperatures also showed dramatic difference between the transgenic plants and wild-type plants.

TPP (trehalose-6-phosphate phosphatase) genes expressed in rice and their expression is induced by cold. Trehalose accumulates rapidly and transiently, which follows the transient induction of TPP activity, in rice tissues during chilling stress. Overexpression of TPS (trehalose-6-phosphate synthase) and TPP genes enhanced the accumulation of trehalose and tolerance to cold stress in transgenic tobacco and rice. However, the regulatory mechanism of TPPs by cold or other stresses is unclear. MYBS3 plays a critical role in cold adaptation in rice and necessary for enhancing cold tolerance. They reported that transgenic rice constitutively overexpressing MYBS3 tolerated 4°C for at least 1 week and exhibited no yield penalty in normal field conditions.

Based on previous studies it has been established that the CBF cold responsive pathway is an integral component of the cold acclimation response. However, the transcriptome data showed that additional cold-regulatory pathways also exist. Transcriptome comparisons indicated that only 12% of the cold-responsive genes are certain members of the CBF regulon. Moreover, at least 28% of the cold-responsive genes were not regulated by the CBF transcription factors, including 15 encoding known or putative transcription factors, indicating that these cold-responsive genes are members of different low-temperature regulons.

When overexpressed in Arabidopsis and tobacco, the soybean gene SCOF-1 (encodes a zinc-finger protein) can activate COR gene expression and increase freezing tolerance in non-acclimated transgenic plants, although the SCOF-1 protein does not directly bind to either the DRE/CRT or the ABRE elements. SCOF-1 interacts with another G-box binding bZIP protein, SGBF-1. SGBF-1 can activate ABRE-driven reporter gene expression in Arabidopsis leaf protoplasts. Thus, SCOF-1 may regulate the activity of SGBF-1 as a transcription factor in inducing COR gene expression.

Forward genetic analysis in Arabidopsis identified two transcription factors, high expression of osmotically responsive genes, HOS9 and HOS10, which are required for basal freezing tolerance. Similarly, microarray analysis led to the identification of the cold-stress-inducible AP2 family transcription factor gene related to ABI3/VP1 (RAV1) that might regulate plant growth under cold stress. RAV1 is down regulated by epibrassinolide and transgenic Arabidopsis overexpressing RAV1 exhibits a retardation of lateral root and rosette-leaf development. Further, the importance of CBF-independent pathways is also supported by analysis of mutants that have increased freezing tolerance, for example, mutations in eskimo1 (ESK1), a protein of unknown function, result in constitutive freezing tolerance. The Eskimo1 mutation was first identified as a mutation conferring frost survival without an acclimation period. Later on, Bouchabke-Coussa et al. clearly showed that ESKIMO1 mutants are more tolerant to freezing but only after acclimation. The genes that are affected by the ESK1 mutation are distinct from those of the CBF regulon. Recently, a CBF 1 gene has been introduced into tomato that resulted in transgenic plants showing higher activity of superoxide dismutase (SOD), higher non-photochemical quenching (NPQ), and lower malondialdehyde (MDA) content. Thus suggesting that CBF1 protein plays an important role in protection of PSII and PSI during low temperature stress at low irradiance.

Aforementioned studies clearly demonstrate the complex and interactive relationships among different pathways regulated by cold acclimation. Hence, elucidation of the mechanism regulating cold-regulated genes is an important goal in achieving a full understanding of cold acclimation. With the advent of molecular genetics and biotechnology, it is now possible to genetically engineer plants to be more tolerant to low temperature. The technologies can be so developed/modified that will significantly improve breeding efficiency, resulting in rapid and accurate incorporation of cold tolerant genes into crop plants.

## Prospects of Cold Tolerant Genes in Hill Agriculture

Cold is an environmental factor that limits the geographical distribution and growing season of many plant species, and it often adversely affects crop quality and productivity.

Most temperate plants can acquire tolerance to freezing temperature by a prior exposure to low non-freezing temperature, a process known as cold acclimatization. Plants of tropical and subtropical origins are sensitive to chilling temperature (0°C-10°C) and are incapable of cold acclimation. Genes induced by stress can be roughly classified into two groups: genes coding for regulatory proteins, mainly transcription factors, and genes encoding proteins involved directly in response mechanisms; genes from both classes are of interest. Variations in the expression of regulators could lead to a protective status before the emergence of stress and have multiple effects. Genes involved in protection or repair mechanisms could be new targets for the improvement of plant plasticity and adaptive responses to stress. The unraveling of general stress responses in the model species Arabidopsis thaliana helped to identify potential targets for plant breeding. Arabidopsis genes involved in tolerance to abiotic stress were transferred, by genetic engineering, to many crops and tolerance was successfully conferred in the field, despite the complexity of plant responses to environmental stress. Thus, finding new key genes responsible for abiotic stress tolerance phenotypes is of great importance not only for a better understanding of stress responses, but also for promising future crop improvement.

Many studies have suggested that cold regulated gene expression is critical in plants for both chilling tolerance and cold acclimation. Cold responsive genes encode a diverse array of proteins such as enzymes involved in respiration and metabolism of carbohydrates, lipids, phenylpropanoids and antioxidants: molecular chaperones, antifreeze proteins, and others with a presumed function in tolerance to dehydration caused by freezing. The change in the gene expression occur in plant during cold acclimatization a developmental process that results in increased tolerance. Since then, it has repeatedly been speculated that certain COR (cold regulated) genes might have role in freezing tolerance. To test this notion investigators have turned to isolating the characterizing genes that are expressed in response to low temperature. These efforts have led to the identification of a number of genes such as the COR 15a KINI, LTI 78, fad 7etc. of A. thaliana. The generic trends of the genes and transcription factors are available in STIFDB (Stress responsive Transcription Factor Database). STIFDB is a database of stress-related genes, which are upregulated in abiotic stress-related microarray experiments. STIFDB provides a platform to understand the stress-regulome of abiotic stress responsive genes in plants. STIFDB will be a highly useful resource for a researcher working on abiotic stress responses in plants.

Low temperature stress is a major environmental factor that not only limits where crops can be grown but also reduces yields depending on the weather in a particular growing season. In addition to exceptionally stressful years that cause significant yield reductions, less extreme stress almost certainly causes smaller losses over large areas to produce comparable yield reductions every year. Even in cases when freezing stress does not result in yield loses; it often results in crop quality reduction. Each year, worldwide losses in crop production due to low temperature damage amount to approximately $ 2 Billion. Some of the major losses include the 1995 early fall frosts in the US which caused losses of over $1 billion to corn and soybeans. The occasional freezes in Florida have shifted the citrus belt further south, and California sustained $650M of damage in 1998 to the citrus crop due to a winter freeze. The inherent cold hardiness of the crop determines in which agricultural areas it can be grown. Crops that are more resistant to freezing stress would allow some geographical regions to grow more profitable and productive crops with less environmental risks. However despite continued efforts, traditional breeding has had only limited success in imparting crop plants with better freezing tolerance due to very little was known about the mechanisms that regulate chilling and freezing tolerance. With the advent of molecular genetics and biotechnology, it is now possible to genetically engineer plants to be more tolerant to many environmental adversities, including low temperature. Molecular studies have shown that several genes with various functions are induced by environmental stresses such as drought, high-salinity and low temperature in plants. Most of the dehydration responsive genes are induced by the plant hormone abscisic acid (ABA), but others are not. Expression analyses of dehydration-responsive genes have provided at least four independent regulatory systems (regulons) for gene expression in a model plant Arabidopsis thaliana. The cis-acting elements in the promoters of some genes that have a typical stress-inducible expression profile and the transcription factors that affect the expression of these genes have been analyzed.

# BREEDING FOR DROUGHT RESISTANCE

Breeding for drought resistance is the process of breeding plants with the goal of reducing the impact of dehydration on plant growth.

## Dehydration Stress

## Crop Plants

In nature or crop fields, water is often the most limiting factor for plant growth. If plants do not receive adequate rainfall or irrigation, the resulting dehydration stress can reduce growth more than all other environmental stresses combined.

Drought can be defined as the absence of rainfall or irrigation for a period of time sufficient to deplete soil moisture and cause dehydration in plant tissues. Dehydration stress results when water loss from the plant exceeds the ability of the plant's roots to absorb water and when the plant's water content is reduced enough to interfere with normal plant processes.

## Global Phenomenon

About 15 million km2 of the land surface is covered by crop-land, and about 16% of this area is equipped for irrigation Thus, in many parts of the world, including the United

States, plants may frequently encounter dehydration stress. Rainfall is very seasonal and periodic drought occurs regularly. The effect of drought is more prominent in sandy soils with low water holding capacity. On such soils some plants may experience dehydration stress after only a few days without water.

During the 20th century, the rate of increase in `blue' water withdrawal (from rivers, lakes, and aquifers) for irrigation and other purposes was higher than the growth rate of the world population.

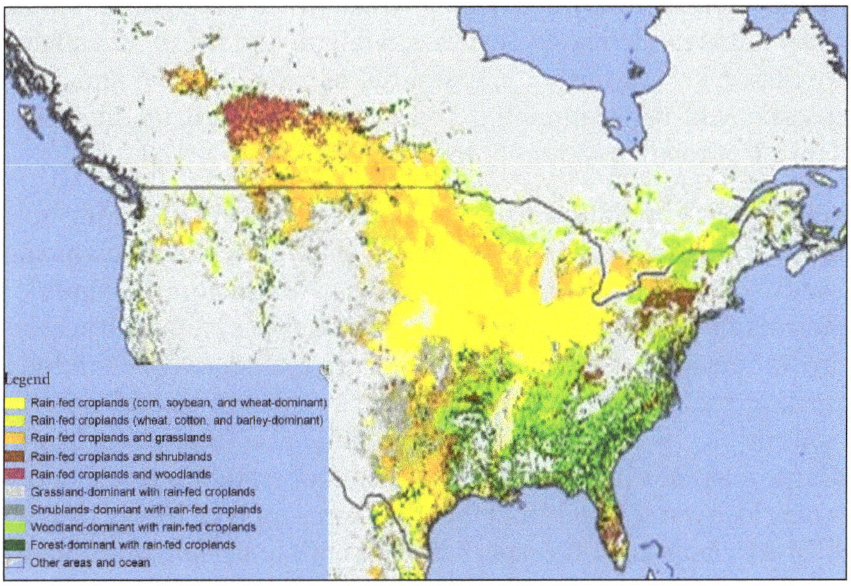


Rain fed areas of USA.

## Future Challenges to Crop Production

Soil moisture deficit is a significant challenge to the future of crop production. Severe drought in parts of the U.S., Australia, and Africa in recent years drastically reduced crop yields and disrupted regional economies. Even in average years, however, many agricultural regions, including the U.S. Great Plains, suffer from chronic soil moisture deficits. Cereal crops typically attain only about 25% of their potential yield due to the effects of environmental stress, with dehydration stress the most important cause. Two major trends will likely increase the frequency and severity of soil moisture deficits:

- Climate change: Higher temperatures are likely to increase crop water use due to increased transpiration. A warmer atmosphere will also speed up melting of mountain snow pack, resulting in less water available for irrigation. More extreme weather patterns will increase the frequency of drought in some regions.
- Limited water supplies: Increased demand from municipal and industrial users will further reduce the amount of water available for irrigated crops.

Although changes in tillage and irrigation practices can improve production by conserving water, enhancing the genetic tolerance of crops to drought stress is considered an essential strategy for addressing moisture deficits.

## Plant Physiology

A plant responds to a lack of water by halting growth and reducing photosynthesis and other plant processes in order to reduce water use. As water loss progresses, leaves of some species may appear to change colour — usually to blue-green. Foliage begins to wilt and, if the plant is not irrigated, leaves will fall off and the plant will eventually die. Soil moisture deficit lowers the water potential of a plant's root and, upon extended exposure, abscisic acid is accumulated and eventually stomatal closure occurs. This reduces a plant's leaf relative water content.

The time required for dehydration stress to occur depends on the water-holding capacity of the soil, environmental conditions, stage of plant growth, and plant species. Plants growing in sandy soils with low water-holding capacity are more susceptible to dehydration stress than plants growing in clay soils. A limited root system will accelerate the rate at which dehydration stress develops. A plant's root system may be limited by the presence of competing root systems from neighbouring plants, by site conditions such as compacted soils or high water tables, or by container size (if growing in a container). A plant with a large mass of leaves in relation to the root system is prone to drought stress as the leaves may lose water faster than the roots can supply it. Newly planted plants and poorly established plants may be especially susceptible to dehydration stress because of the limited root system or the large mass of stems and leaves in comparison to roots.

## Other Stress Factors

Aside from the moisture content of the soil, environmental conditions of high light intensity, high temperature, low relative humidity and high wind speed will significantly increase plant water loss. The prior environment of a plant also can influence the development of dehydration stress. A plant that has been exposed to dehydration stress (hardened) previously and has recovered may become more drought resistant. Also, a plant that was well-watered prior to being water-limited will usually survive a period of drought better than a continuously dehydration-stressed plant.

## Mechanisms of Drought Resistance

The degree of resistance to drought depends upon individual crops. Generally three strategies can help a crop to mitigate the effect of dehydration stress:

## Avoidance

If the plant shows dehydration avoidance, the environmental factor is excluded from the plant tissues by reducing water loss ("water savers", e.g. by thick leaf epicuticular wax, leaf rolling, leaf posture) or maintaining water uptake ("water spenders", e.g. by growing deeper roots). Dehydration avoidance is desirable in modern agriculture, where drought resistance requires the maintenance of economically viable plant production under dehydration stress. The role of dehydration avoidance is maintaining water supply and sustaining leaf hydration and turgidity with the purpose of maintaining stomatal opening and transpiration as long as possible under water deficit. This is essential for leaf gas exchange, photosynthesis and plant production through carbon assimilation.

## Tolerance

If the plant shows dehydration tolerance, the environmental factor enters the plant tissues but the tissues survive, by e.g. maintaining turgor and osmotic adjustment.

## Escape

Dehydration escape involves e.g. early maturing or seed dormancy, where the plant uses previous optimal conditions to develop vigor. Dehydration recovery refers to some plant species being able to recuperate after brief drought periods.

A proper timing of life-cycle, resulting in the completion of the most sensitive developmental stages while water is abundant, is considered to be a dehydration escape strategy. Avoiding dehydration stress with a root system capable of extracting water from deep soil layers, or by reducing evapotranspiration without affecting yields, is considered as dehydration avoidance. Mechanisms such as osmotic adjustment (OA) whereby a plant maintains cell turgor pressure under reduced soil water potential are categorised as dehydration tolerance mechanisms. Dehydration avoidance mechanisms can be expressed even in the absence of stress and are then considered constitutive. Dehydration tolerance mechanisms are the result of a response triggered by dehydration stress itself and are therefore considered adaptive. When the stress is terminal and predictable, dehydration escape through the use of shorter duration varieties is often the preferable method of improving yield potential. Dehydration avoidance and tolerance mechanisms are required in situations where the timing of drought is mostly unpredictable.

Drought resistance mechanisms are genetically controlled and genes or QTL responsible for drought resistance have been discovered in several crops which opens avenue for molecular breeding for drought resistance.

## Drought Resistance Traits

Resistance to drought is a quantitative trait, with a complex phenotype, often confounded by plant phenology. Breeding for drought resistance is further complicated since several types of abiotic stress, such as high temperatures, high irradiance, and nutrient toxicities or deficiencies can challenge crop plants simultaneously.

## Osmotic Adjustment

When a plant is exposed to water deficit, it may accumulate a variety of osmotically active compounds such as amino acids and sugars, resulting in a lowering of the osmotic potential. Examples of amino acids that may be up-regulated are proline and glycine betaine. This is termed osmotic adjustment and enables the plant to take up water, maintain turgor and survive longer.

## Cell Membrane Stability

The ability to survive dehydration is influenced by a cell's ability to survive at reduced water content. This can be considered complementary to OA because both traits will help maintain leaf growth (or prevent leaf death) during water deficit. Crop varieties differ in dehydration tolerance and an important factor for such differences is the capacity of the cell membrane to prevent electrolyte leakage at decreasing water content, or "cell membrane stability (CMS)". The maintenance of membrane function is assumed to mean that cell activity is also maintained. Measurements of CMS have been used in different crops and are known to be correlated with yields under high temperature and possibly under dehydration stress.

## Epicuticular Wax

In sorghum, drought resistance is a trait that is highly correlated with the thickness of the epicuticular wax layer. Experiments have demonstrated that rice varieties with a thick cuticle layer retain their leaf turgor for longer periods of time after the onset of a water-stress.

## Partitioning and Stem Reserve Mobilisation

As photosynthesis becomes inhibited by dehydration, the grain filling process becomes increasingly reliant on stem reserve utilisation. Numerous studies have reported that stem reserve mobilisation capacity is related to yield under dehydration stress in wheat. In rice, a few studies also indicated that this mechanism maintains grain yield under dehydration stress at the grain filling stage. This dehydration tolerance mechanism is stimulated by a decrease in gibberellic acid concentration and an increase in abscisic acid concentration.

## Manupulation and Stability of Flowering Processes

## Seedling Drought Resistance Traits

For emergence from deep sowing (to exploit dry upper soil), this is practised to help seedlings reach the receding moisture profile, and to avoid high soil surface temperatures which inhibit germination. Screening at these stage provides practical advantages, specially when managing large amount of germ-plasms.

## The Drought Resistant Ideotype

Usually ideotypes are developed to create an ideal plant variety. The following traits constitute ideotype of wheat by CIMMYT:

- Large seed size. Helps emergence, early ground cover and initial biomass.
- Long coleoptiles. For emergence from deep sowing.
- Early ground cover.

Thinner, wider leaves (i.e., with a relatively low specific leaf weight) and a more prostrate growth habit help to increase ground cover, thus conserving soil moisture and potentially increasing radiation use efficiency:

- High pre-anthesis biomass.
- Good capacity for stem reserves and remobilisation.
- High spike photosynthetic capacity.
- High RLWC/Gs/CTD during grain filling to indicate ability to extract water.
- Osmotic adjustment.
- Accumulation of ABA.

The benefit of ABA accumulation under dehydration has been demonstrated. It appears to pre-adapt plants to stress by reducing stomatal conductance, rates of cell division, organ size, and increasing development rate. However, high ABA can also result in sterility problems since high ABA levels may abort developing florets:

- Heat Tolerance: The contribution of heat tolerance to performance under dehydration stress needs to be quantified, but it is relatively easy to screen for.
- Leaf anatomy: Waxiness, pubescence, rolling, thickness, posture. These traits decrease radiation load to the leaf surface. Benefits include a lower evapotranspiration rate and reduced risk of irreversible photo-inhibition. However, they may also be associated with reduce radiation use efficiency, which would reduce yield under more favourable conditions.
- High tiller survival: Comparison of old and new varieties have shown that under dehydration older varieties over-produce tillers many of which fail to set grain while modern drought resistant lines produce fewer tillers most of which survive.
- Stay-green: The trait may indicate the presence of drought resistance mechanisms, but probably does not contribute to yield per se if there is no water left in the soil profile by the end of the cycle to support leaf gas exchange. It may be detrimental if it indicates lack of ability to remobilise stem reserves. However, research in sorghum has indicated that Stay-green is associated with higher leaf

chlorophyll content at all stages of development and both were associated with improved yield and transpiration efficiency under dehydration.

### Combination Phenomics: Overall Health of Crops

The concept of combination phenomics comes from the idea that two or more plant stresses have common physiological effects or common traits - which are an indicator of overall plant health. As both biotic and abiotic stresses can result in similar physiological consequence, drought resistant plants can be separated from sensitive plants. Some imaging or infrared measuring techniques can help to speed the process for breeding process. For example, spot blotch intensity and canopy temperature depression can be monitored with canopy temperature depression.

## BREEDING FOR HEAT STRESS TOLERANCE

Plant breeding is process of development of new cultivars. Plant breeding involves development of varieties for different environmental conditions – some of them are not favorable. Among them, heat stress is one of such factor that reduces the production and quality significantly. So breeding against heat is a very important criterion for breeding for current as well as future environments produced by global climate change (e.g. global warming).

### Breeding for Heat Stress Tolerance in Plants

Heat stress due to increased temperature is a very important problem globally. Occasional or prolonged high temperatures cause different morpho-anatomical, physiological and biochemical changes in plants. The ultimate effect is on plant growth as well as development and reduced yield and quality. Breeding for heat stress tolerance can be mitigated by breeding plant varieties that have improved levels of thermo-tolerance using different conventional or advanced genetic tools. Marker assisted selection techniques for breeding are highly useful. Recently 41 polymorphic SSR markers has been identified between a heat tolerant rice variety 'N22' and heat susceptible-high yielding variety 'Uma' for the development of new 'high yielding-heat tolerant' rice varieties.

Heat stress is defined as increased temperature level sufficient to cause irreversible damage to plant growth and development. Generally a temperature rise, above usually 10 to 15°C above ambient, can be considered heat shock or heat stress. Heat tolerance is broadly defined as the ability of the plant tolerate heat – means that grow and produce economic yield under high temperatures.

### Significance: Current and Future

Heat stress is a serious threat to crop production globally. Global warming is particularly consequence of increased level of greenhouse gases such as $CO_2$, methane, chlorofluorocarbons and nitrous oxides. The Intergovernmental Panel on Climatic Change (IPCC) has predicted a rise of 0.3°C per decade reaching to approximately 1 and 3°C above the present value by 2025 and 2100 AD, respectively.

## Physiological Consequence of Heat Stress

At very high temperatures cause severe cellular injury and cell death may occur within short time, thus leading to a catastrophic collapse of cellular organization. However, under moderately high temperatures, the injury can only occur after longer exposure to such a temperature however the plant efficiency can be severely affected. High temperature directly affects injuries such as protein denaturation and aggregation, and increased fluidity of membrane lipids. Other indirect or slower heat injuries involve inactivation of enzymes in chloroplast and mitochondria, protein degradation, inhibition of protein synthesis, and loss of membrane integrity. Heat stress associated injuries ultimately lead to starvation, inhibition of growth, reduced ion flux, production of toxic compounds and production of reactive oxygen species (ROS). Immediately after exposure to high temperature stress-related proteins are expressed as stress defense strategy of the cell. Expression of heat shock proteins (HSPs), protein with 10 to 200 kDa, is supposed to be involved in signal transduction during heat stress. In many species it has been demonstrated that HSPs results in improved physiological phenomena such as photosynthesis, assimilate partitioning, water and nutrient use efficiency, and membrane stability.

Studies have found tremendous variation within and between species, thus this will help to breed heat tolerance for future environment. Some of attempts to develop heat-tolerant genotypes are successful.

## Traits Associated with Heat Stress Tolerance

Different physiological mechanisms may contribute to heat tolerance in the field—for example, heat tolerant metabolism as indicated by higher photosynthetic rates, stay-green, and membrane thermo-stability, or heat avoidance as indicated by canopy temperature depression. Several physiological and morphological traits have been evaluated for heat tolerance - Canopy temperature, leaf chlorophyll, stay green, leaf conductance, spike number, biomass, and flowering date.

## Canopy Temperature Depression (CTD)

CTD has shown clear association with yield in warm environments shows it association with heat stress tolerance. CTD shows high genetic correlation with yield and high values of proportion of direct response to selection indicating that the trait is heritable and therefore amenable to early generation selection. Since an integrated CTD value can be measured almost instantaneously on scores of plants in a small breeding plot (thus reducing error normally associated with traits measured on individual plants), work has been conducted to evaluate its potential as an indirect selection criterion for genetic gains in yield. CTD is affected by many physiological factors, which makes it a powerful.

## Stomatal Conductance

Canopy temperature depression is highly suitable for selecting physiologically superior lines in warm, low relative humidity environments where high evaporative demand leads to leaf cooling of up to 10°C below ambient temperatures. This permits differences among genotypes to be detected relatively easily using infrared thermometry. However, such differences cannot be detected in high relative humidity environments because the effect of evaporative cooling of leaves is negligible. Nonetheless, leaves maintain their stomata open to permit the uptake of $CO_2$, and differences in the rate of $CO_2$ fixation may lead to differences in leaf conductance that can be measured using a porometer. Porometry can be used to screen individual plants. The heritability of stomatal conductance is reasonably high, with reported values typically in the range of 0.5 to 0.8. Plants can be assessed for leaf conductance using a viscous flow porometer that is available on the market. This instrument can give a relative measure of stomatal conductance in a few seconds, making it possible to identify physiologically superior genotypes from within bulks.

## Membrane Thermostability

Although resistance to high temperatures involves several complex tolerance and avoidance mechanisms, the membrane is thought to be a site of primary physiological injury by heat, and measurement of solute leakage from tissue can be used to estimate damage to membranes. Since membrane thermostability is reasonably heritable and shows high genetic correlation with yield.

## Chlorophyll Fluorescence

Chlorophyll fluorescence, an indication of the fate of excitation energy in the photosynthetic apparatus, has been used indicator for heat stress tolerance.

## Chlorophyll Content and Stay Green

Chlorophyll content and stay green traits have been found to be associated with heat stress tolerance. Xu et al. identified three QTLs for chlorophyll content (Chl1, Chl2, and Chl3) (coincided with three stay-green QTL regions (Stg1, Stg2, and Stg3)) were identified in Sorghum. The Stg1 and Stg2 regions also contain the genes for key photosynthetic enzymes, heat shock proteins, and an abscisic acid (ABA) responsive gene.

## Photosynthesis

Decliend photosynthesis is suggested as measure of heat stress sensitivity in plants.

## Combination Breeding and Physiological Breeding

The physiological-trait-based breeding approach has merit over breeding for yield *per se* because it increases the probability of crosses resulting in additive gene action. The concept of combination phenomics comes from the idea that two or more stress have common physiological effect or common traits - which is an indicator of overall plant health. Similar analogy in human medical terms is high blood pressure or high body temperature or high white blood cells in body is an indicator of health problems and thus wecan select healthy people from unhealthy using such a measure. As both abiotic and abiotic stresses can result in similar physiological consequence, tolerant plant can be separated from sensitive plants. Some imaging or infrared measuring techniques can help to speed the process for breeding process. For example, spot blotch intensity and canopy temperature depression can be monitored with canopy temperature depression.

# PERMISSIONS

All chapters in this book are published with permission under the Creative Commons Attribution Share Alike License or equivalent. Every chapter published in this book has been scrutinized by our experts. Their significance has been extensively debated. The topics covered herein carry significant information for a comprehensive understanding. They may even be implemented as practical applications or may be referred to as a beginning point for further studies.

We would like to thank the editorial team for lending their expertise to make the book truly unique. They have played a crucial role in the development of this book. Without their invaluable contributions this book wouldn't have been possible. They have made vital efforts to compile up to date information on the varied aspects of this subject to make this book a valuable addition to the collection of many professionals and students.

This book was conceptualized with the vision of imparting up-to-date and integrated information in this field. To ensure the same, a matchless editorial board was set up. Every individual on the board went through rigorous rounds of assessment to prove their worth. After which they invested a large part of their time researching and compiling the most relevant data for our readers.

The editorial board has been involved in producing this book since its inception. They have spent rigorous hours researching and exploring the diverse topics which have resulted in the successful publishing of this book. They have passed on their knowledge of decades through this book. To expedite this challenging task, the publisher supported the team at every step. A small team of assistant editors was also appointed to further simplify the editing procedure and attain best results for the readers.

Apart from the editorial board, the designing team has also invested a significant amount of their time in understanding the subject and creating the most relevant covers. They scrutinized every image to scout for the most suitable representation of the subject and create an appropriate cover for the book.

The publishing team has been an ardent support to the editorial, designing and production team. Their endless efforts to recruit the best for this project, has resulted in the accomplishment of this book. They are a veteran in the field of academics and their pool of knowledge is as vast as their experience in printing. Their expertise and guidance has proved useful at every step. Their uncompromising quality standards have made this book an exceptional effort. Their encouragement from time to time has been an inspiration for everyone.

The publisher and the editorial board hope that this book will prove to be a valuable piece of knowledge for students, practitioners and scholars across the globe.

# INDEX